Encyclopaedic Dictionary of Telecommunication

Vol. 3

Rajiv Shukla

RAJAT PUBLICATIONS

NEW DELHI - 110 002 (INDIA)

RAJAT PUBLICATIONS
4740/23, Ansari Road, Daryaganj
New Delhi- 110 002 (India)
Phones : +91-11-23267924, +91-9811484973
E-mail: rajat_publication@yahoo.com

Encyclopaedic Dictionary of Telecommunication

Edition 2025

ISBN 978-81-7880-307-4 (Set)

[The responsibility for facts stated opinion expressed or conclusions reached and plagiarism, if any, in this volume is entirely that of the Editor. The publisher bears no responsibility for them whatsoever.]

PRINTED IN INDIA

Published by Mrs. Seema Wasan for Rajat Publications, New Delhi and Printed at H.S. Offset Press, Delhi.

Preface

Telecommunications is the transmission of data and information between computers using a communications link such as a standard telephone line. Typically, a basic telecommunications system would consist of a computer or terminal on each end, communication equipment for sending and receiving data, and a communication channel connecting the two users. Appropriate communications software is also necessary to manage the transmission of data between computers. Some applications that rely on this communications technology include the following:

1. Electronic mail (e-mail) is a message transmitted from one person to another through computerized channels. Both the sender and receiver must have access to on-line services if they are not connected to the same network. E-mail is now one of the most frequently used types of telecommunication.

2. Facsimile (fax) equipment transmits a digitized exact image of a document over telephone lines. At the receiving end, the fax machine converts the digitized data back into its original form.

3. Voice mail is similar to an answering machine in that it permits a caller to leave a voice message in a voice mailbox. Messages are digitized so the caller's message can be stored on a disk.

4. Videoconferencing involves the use of computers, television cameras, and communications software and equipment. This equipment makes it possible to conduct electronic meetings while the participants are at different locations.

5. The Internet is a continuously evolving global network of computer networks that facilitates access to information on thousands of topics. The Internet is utilized by millions of people daily.

Actually, telecommunications is not a new concept. It began in the mid-1800s with the telegraph, whereby sounds were translated manually into words; then the telephone, developed in 1876, transmitted voices; and then the teletypewriter, developed in the early 1900s, was able to transmit the written word.

Since the 1960s, telecommunications development has been rapid and wide reaching. The development of dial modem technology accelerated the rate during the 1980s. Facsimile transmission also enjoyed rapid growth during this time. The 1990s have seen the greatest advancement in telecommunications. It is predicted that computing performance will double every eighteen months. In addition, it has been estimated that the power of the computer has doubled thirty-two times since World War II (With row, 1997). The rate of advancement in computer technology shows no signs of slowing.

Deregulation and new technology have created increased competition and widened the range of network services available throughout the world. This increase in telecommunication capabilities allows businesses to benefit from the information revolution in numerous ways, such as streamlining their inventories, increasing productivity, and identifying new markets.

True source for the latest terms and concepts used today in the field of telecommunications. The objective of this book is to provide a self-contained quick-reference to telecommunications jargon and facts in a clear concise manner.

Written and reviewed by industry expert, this dictionary cover topics spanning the telecommunications and information technology industry, including distribution design, insatllation, outside plant, and LAN and internetworking design. The dictionary is based on recognized codes, standards, and regulations, and include extensive glossaries.

—Editor

PABX. *Abbreviation for* private automatic branch exchange. *See* PBX. *Note:* Use of the term *"PBX"* is more common than *"PABX,"* regardless of automation.

packet. In data communication, a sequence of binary digits, including data and control signals, that is transmitted and switched as a composite whole. *Note:* The data, control signals, and possibly error control information, are arranged in a specific format.

packet assembler/disassembler (PAD). A functional unit that enables data terminal equipment (DTE) not equipped for packet switching to access a packet-switched network.

packet format. The structure of data, address, and control information in a packet. *Note:* The size and content of the various fields in a packet are defined by a set of rules that are used to assemble the packet.

packet Internet groper. *See* ping.

packet mode. A mode of operating a communications network in which packet switching is used rather than message switching.

packet-mode terminal. Data terminal equipment (DTE) that can control, format, transmit, and receive packets.

packet-switched data transmission service. A service that (a) provides for the transmission of data in the form of packets, (b) switches data at the packet level, and (c) may provide for the assembly and disassembly of data packets.

packet switching. The process of routing and transferring data by means of addressed packets so that a channel is occupied during the transmission of the packet only, and upon completion of the transmission the channel is made available for the transfer of other traffic.

packet-switching network. A switched network that transmits data in the form of packets.

packet-switching node. In a packet-switching network, a node that contains data switches and equipment for controlling, formatting, transmitting, routing, and receiving data packets. *Note:* In the Defense Data Network (DDN), a packet-switching node is usually configured to support up to thirty-two X.25 56-kb/s host connections, as many as six 56-kb/s interswitch trunk (IST) lines to other packet-switching nodes, and at least one Terminal Access Controller (TAC).

packet transfer mode. A method of information transfer, by means of packet transmission and packet switching, that permits dynamic sharing of network resources among many connections.

packing density. The number of storage cells per unit length, area, or volume of storage media. *Note:* Examples of packing density are the number of bits or characters stored per unit length of magnetic tape and the number of bits stored per unit length of track on an optical disk.

pad. A network, of fixed resistors, that attenuates signals by a fixed amount with negligible distortion. *Note:*

The resistive network is called an *attenuator* if the resistance is adjustable.

PAD. *Acronym for* packet assembler/disassembler.

pager. A mobile receiver for paging communications, also known as a *"beeper."*

paging. A one-way communications service from a base station to mobile or fixed receivers that provide signaling or information transfer by such means as tone, tone-voice, tactile, optical readout, *etc.*

paired cable. A cable made up of one or more separately insulated twisted-wire pairs, none of which is arranged with another to form quads.

paired disparity code. A code in which some or all of the characters are represented by two sets of digits of opposite disparity that are used in sequence so as to minimize the total disparity of a longer sequence of digits. *Note 1:* An alternate mark inversion signal is an implementation of a paired disparity code. *Note 2:* The digits may be represented by disparate physical quantities, such as two different frequencies, phases, voltage levels, magnetic polarities, or electrical polarities, each one of the pair representing a 0 or a 1.

pair-gain system. A transmission system that uses concentrators or multiplexers so that fewer wire pairs may be used than would otherwise be required to provide service to a given number of subscribers.

PAL. *Acronym for* phase alternation by line. A television signal standard (625 lines, 50 Hz, 220 V primary power) used in the United Kingdom, much of the rest of western Europe, several South American countries, some Middle East and Asian countries, several African countries, Australia, New Zealand, and other Pacific island countries.

PAL-M. A modified version of the phase-alternation-by-line (PAL) television signal standard (525 lines, 50 Hz, 220 V primary power), used in Brazil.

PAM. *Abbreviation for* pulse-amplitude modulation.

PAMA. *Abbreviation for* pulse-address multiple access.

panning. 1. On the viewing screen of a display device, *e.g.,* a computer monitor, horizontal shifting of the entire displayed image. *Note:* The panning direction is at a right angle with respect to the scrolling direction. 2. In video technology, the use of a camera to scan a subject horizontally. 3. In antenna systems, successively changing the azimuth of a beam of radio-frequency energy over the elements of a given horizontal region, or the corresponding process in reception.

***p/a* r.** *Abbreviation for* peak-to-average ratio.

parabolic antenna. An antenna consisting of a parabolic reflector and a radiating or receiving element at or near its focus. *Note:* If the reflector is in the shape of a paraboloid of revolution, it is called a paraboloidal reflector; cylindrical paraboloids and off-axis paraboloids of revolution are also used.

parabolic profile. In an optical fiber, a power-law index profile with the profile parameter, g, equal to 2. *Synonym* quadratic profile.

parallel computer. A computer that has multiple arithmetic units or logic units that are used to accomplish parallel operations or parallel processing.

parallel port. A port through which two or more data bits are passed simultaneously, such as all the bits of an 8-bit byte, and that requires as many input channels as the number of bits that are to be handled simultaneously. *Contrast with* serial port.

parallel processing. Pertaining to the concurrent or simultaneous execution of two or more processes in a single unit.

parallel-to-serial conversion. Conversion of a stream of multiple data elements, received simultaneously, into a stream of data elements transmitted in time sequence, *i.e.*, one at a time. *Contrast with* serial-to-parallel conversion.

parallel transmission. 1. The simultaneous transmission of the signal elements of a character or other data item. 2. In digital communications, the simultaneous transmission of related signal elements over two or more separate paths. *Note:* Protocols for parallel transmission, such as those used for computer ports, have been standardized by ANSI.

parametric amplifier (paramp). An amplifier that (a) has a very low noise level, (b) has a main oscillator that is tuned to the received frequency, (c) has another pumping oscillator of a different frequency that periodically varies the parameters, *i.e.,* the capacitance or inductance, of the main oscillator circuit, and (d) enables amplification of the applied signal by making use of the energy from the pumping action. *Note:* Paramps with a variable-capacitance main-oscillator semiconductor diode are used in radar tracking and communications Earth stations, Earth satellite stations, and deep-space stations. The noise temperature of paramps cooled to the temperature of liquid helium, about 5 K, is in the range of 20 to 30 K. Paramp gains are about 40 dB.

parasitic element. Of an antenna, a directive element that is not connected to a radio transmitter or receiver either directly or via a feeder, but is coupled to the driven element only by the fields. *Synonym* passive element.

parasitic emission. In a communications system in which one or more electromagnetic sources are used, electromagnetic radiation–such as lightwaves, radio waves, microwaves, X-rays, or gamma rays from one or more of the sources–that is not harmonically related, *i.e.,* is not coherent, with the transmitted carrier. *Note:* Parasitic emissions are usually caused by undesired oscillations or energy-level transitions in the sources.

paraxial ray. In optical systems, a ray that is close to and nearly parallel with the optical axis.

parity. In binary-coded data, a condition that is maintained such that, in any permissible coded expression, the total number of 1s, or 0s, is always odd or always even. *Note 1:* Parity is used in error-detecting and error-correcting codes. *Note 2:* For example, in the ASCII code or in the International Telegraph Alphabet 5 (ITA-5) code as usually implemented, 7 bits are used to represent each character and 1 bit is used as a parity check bit.

parity check. A test that determines whether the number of ones or zeros in an array of binary digits is odd or even. *Note:* Odd parity is standard for synchronous transmission and even parity for asynchronous transmission. *Synonym* odd-even check.

par meter. *Abbreviation for* peak-to-average ratio meter. A meter used to measure, calculate, and display the ratio of the peak power level to the time-averaged power level in a circuit, *i.e.*, the peak-to-average ratio (*p/a* r). *Note 1:* A par meter is used as a quick means to identify degraded telephone channels. *Note 2:* A par meter is very sensitive to envelope delay distortion. The par meter may also be used for idle channel noise, nonlinear distortion, and amplitude-distortion measurements. *Note 3:* The peak-to-average ratio can be determined for many signal parameters, such as voltage, current, power, frequency, and phase.

part 68. The section of Title 47 of the *Code of Federal Regulations* governing (a) the direct connection of telecommunications equipment and customer premises wiring with the public switched telephone network and certain private line services, such as (1) foreign exchange lines at the customer premises end, (2) the station end of off-premises stations associated with PBX and Centrex® services, (3) trunk-to-station tie lines at the trunk end only, and (4) switched service network station lines, *i.e.*, common control switching arrangements; and (b) the direct connection of (1) all PBX and similar systems to private line services for tie trunk type interfaces, (2) off-premises station lines, and (3) automatic identified outward dialing and message registration. *Note:* Part 68 rules provide the technical and procedural standards under which direct electrical connection of customer-provided telephone equipment, systems, and protective apparatus may be made to the nationwide network without causing harm and without a requirement for protective circuit arrangements in the service-provider networks.

party line. In telephone systems, an arrangement in which two or more user end instruments, usually telephones, are connected to the same loop. *Note:* If selective ringing is not used, individual users may be alerted by different ringing signals, such as a different number of rings or a different combination of long and short rings. Party lines remain primarily in rural areas where loops are long. Privacy is limited and congestion often occurs. *Synonym* multiparty line.

passband. The portion of spectrum, between limiting frequencies (or, in the optical regime, limiting wavelengths), that is transmitted with minimum relative loss or maximum relative gain. *Note 1:* The limiting frequencies are defined as those at which the relative intensity or power decreases to a specified fraction of the maximum intensity or power. This decrease in power is often specified to be the half-power points, *i.e.*, 3 dB below the maximum power. *Note 2:* The difference between the limiting frequencies is called the bandwidth, and is expressed in hertz (in the optical regime, in nanometers or micrometers).

passive device. A device that does not require a source of energy for its operation. *Note:* Examples of passive devices are electrical resistors, electrical capacitors, diodes, optical fibers, cables, wires, glass lenses, and filters.

passive element. *Synonym* parasitic element.

passive network. A network that does not require a power source for its operation.

passive repeater. An unpowered device used to route a microwave beam over or around an obstruction. *Note:* Examples of passive repeaters are (a) two parabolic antennas connected back-to-back, and (b) a flat reflector used as a mirror.

passive satellite. In a satellite communications system, a satellite that only reflects signals from one Earth station to another, or from several Earth stations to several others. *Note:* Although the satellite acts passively by reflecting signals, it may contain active devices for station keeping.

passive sensor. A measuring instrument in the Earth exploration-satellite service or in the space research service by means of which information is obtained by reception of radio waves of natural origin.

passive star. *See* star coupler.

passive station. On a multipoint connection or a point-to-point connection using basic mode link control, any tributary station waiting to be polled or selected.

password. 1. [A] protected/private character string used to authenticate an identity or to authorize access to data. 2. In data communications, a word, character, or combination thereof, that permits access to otherwise inaccessible data, information, or facilities.

password length equation. An equation that determines an appropriate password length, M, which provides an acceptable probability, P, that a password will be guessed in its lifetime. *Note:* The password length is given by $M = (\log S)/(\log N)$ where S is the size of the password space and N is the number of characters available. The password space is given by $S = LR/P$, where L is the maximum lifetime of a password and R is the number of guesses per unit of time.

password length parameter. A basic parameter affecting the password length needed to provide a given degree of security. *Note 1:* Password length parameters are related by the expression $P = LR/S$, where P is the probability that a password can be guessed in its lifetime, L is the maximum lifetime a password can be used to log in to a system, R is the number of guesses per unit of time, and S is the number of unique algorithm-generated passwords (the password space). *Note 2:* The degree of password security is determined by the probability that a password can be guessed in its lifetime.

patch. 1. To connect circuits together temporarily. *Note:* In communications, patches may be made by means of a cord, *i.e.,* a cable, known as a "patch cord." In automated systems, patches may be made electronically. 2. In a computer program, one or more statements inserted to circumvent a problem or to alter temporarily or permanently a usually limited aspect or characteristic of the functioning of the program, *e.g.*, to customize the program for a particular application or environment.

patch and test facility (PTF). A facility in which supporting functions, such as (a) quality control checking and testing of equipment, links, and circuits, (b) troubleshooting, (c) activating, changing, and deactivating of circuits, and (d) technical coordinating and reporting, are performed.

patch bay. An assembly of hardware so arranged that a number of circuits, usually of the same or similar type, appear on jacks for monitoring, interconnecting, and testing purposes. *Note 1:* Patch bays are used at many locations, such as technical control facilities, patch and test facilities, and at telephone exchanges. *Note 2:* Patch bays facilitate flexibility in the use, routing or restoration of a variety of circuit types, such as dc, VF, group, coaxial, equal-level, and digital data circuits.

patch panel. One segment of a patch bay.

path. 1. In communications systems and network topologies, a route between any two points. 2. In radio communications, the route that (a) lies between a transmitter and a receiver and (b) may consist of two or more concatenated links. *Note:* Examples of paths are line-of-sight paths and ionospheric paths. 3. In a computer program, the logical sequence of instructions executed by a computer. 4. In database management systems, a series of physical or logical connections between records or segments, usually requiring the use of pointers.

path attenuation. *Synonym* path loss.

path clearance. In microwave line-of-sight communications, the perpendicular distance from the radio-beam axis to obstructions such as trees, buildings, or terrain. *Note:* The required path clearance is usually expressed, for a particular *k*-factor, as some fraction of the first Fresnel zone radius.

path intermodulation noise. *See* intermodulation noise.

path loss. In a communication system, the attenuation undergone by an electromagnetic wave in transit between a transmitter and a receiver. *Note 1:* Path loss may be due to many effects such as free-space loss, refraction, reflection, aperture-medium coupling loss, and absorption. *Note 2:* Path loss is usually expressed in dB. *Synonym* path attenuation.

path profile. A graphic representation of the physical features of a propagation path in the vertical plane containing both endpoints of the path, showing the surface of the Earth and including trees, buildings, and other features that may obstruct the radio signal. *Note:* Profiles are drawn either with an effective Earth radius simulated by a parabolic arc—in which case the ray paths are drawn as straight lines—or with a *"flat Earth"*—in which case the ray paths are drawn as parabolic arcs.

path quality analysis. In a communications path, an analysis that (a) includes the overall evaluation of the component quality measures, the individual link quality measures, and the aggregate path quality measures, and (b) is performed by evaluating communications parameters, such as bit error ratio, signal-plus-noise-plus-distortion to noise-plus-distortion ratio, and spectral distortion.

path quality matrix. A data bank that contains path-quality analyses used to support path selection and routing determination. *Note:* In adaptive radio automatic link establishment, path quality matrices contain path quality data for single-link and multilink paths.

path survey. The assembling of pertinent geographical and environmental data required to design a radio communication system.

pattern recognition. The identification of objects and images by their shapes, forms, outlines, color, surface texture, temperature, or other attribute, usually by automatic means.

Pawsey stub. A device for connecting an unbalanced coaxial feeder to a balanced antenna.

PAX. *Abbreviation for* private automatic exchange. *See* PBX.

payload. In a set of data, such as a data field, block, or stream, being processed or transported, the part that represents user information and user overhead information, and may include user-requested additional information, such as network management and accounting information. *Note:* The payload does not include system overhead information for the processing or transportation system. *Synonym* mission bit stream.

payload module. The portion of a payload that completely occupies one or more channels.

PBER. *Abbreviation for* pseudo bit-error ratio. In adaptive high-frequency (HF) radio, a bit error ratio derived by a majority decoder that processes redundant transmissions. *Note:* In adaptive HF radio automatic link establishment, PBER is determined by the extent of error correction, such as by using the fraction of non-unanimous votes in the 2-of-3 majority decoder.

PBX. *Abbreviation for* private branch exchange. 1. A subscriber-owned telecommunications exchange that usually includes access to the public switched network. 2. A switch that serves a selected group of users and that is subordinate to a switch at a higher level military establishment. 3. A private telephone switchboard that provides on-premises dial service and may provide connections to local and trunked communications networks. *Note 1:* A PBX operates with only a manual switchboard; a private automatic exchange (PAX) does not have a switchboard, a private automatic branch exchange (PABX) may or may not have a switchboard. *Note 2:* Use of the term *"PBX"* is far more common than *"PABX,"* regardless of automation.

PBX tie trunk: *See* tie trunk.

PBX trunk. *See* trunk.

PC. *Abbreviation for* carrier power (of a radio transmitter).

PCB. *Abbreviation for* power circuit breaker.

PCM. *Abbreviation for* pulse-code modulation.

PCM multiplex equipment. *See* multiplexer.

PCS. *Abbreviation for* Personal Communications Service. A set of capabilities that allows some combination of terminal mobility, personal mobility, and service profile management. *Note 1:* The flexibility offered by PCS can supplement existing telecommunications services, such as cellular radio, used for NS/EP missions. *Note 2:* PCS and UPT are sometimes mistakenly assumed to be the same service concept. UPT allows complete personal mobility across multiple networks and service providers. PCS may use UPT concepts to improve subscriber mobility in allowing roaming to different service providers, but UPT and PCS are not the same service concept. *Contrast with* Universal Personal Telecommunications service.

PCS switching center. In personal communications service, a facility that (a) supports access-independent call control/service control, and connection control (switching) functions, and (b) is responsible for interconnection of access and network systems to support end-to-end services. *Note 1:* The PCS switching center represents a collection of one or more network elements. *Note 2:* The term *"center"* does not imply a physical location.

PCS System. In personal communications service, a collection of facilities that provides some combination of personal mobility, terminal mobility, and service profile management. *Note:* As used here, "facilities" includes hardware, software, and network components such as transmission facilities, switching facilities, signaling facilities, and databases.

PDM. *Abbreviation for* pulse-duration modulation.

PDN. *Abbreviation for* public data network.

PDS. *Abbreviation for* protected distribution system.

PDU. *Abbreviation for* protocol data unit.

PE. *Abbreviation for* phase-encoded. *See* phase-encoded recording.

peak busy hour. *Synonym* busy hour.

peak emission wavelength. Of an optical emitter, the spectral line having the greatest power. *Synonym* peak wavelength.

peak envelope power (of a radio transmitter) [PEP, pX, PX]. The average power supplied to the antenna transmission line by a transmitter during one radio frequency cycle at the crest of the modulation envelope taken under normal operating conditions.

peak limiting. A process by which the absolute instantaneous value of a signal parameter is prevented from exceeding a specified value.

peak power output. The output power averaged over that cycle of an electromagnetic wave having the maximum peak value that can occur during transmission.

peak signal level. 1. In a transmission path, the maximum instantaneous signal power, voltage, or current at any point. 2. At a given point in a transmission path, the maximum instantaneous signal power, voltage, or current that occurs during a specified period.

peak spectral emission. *See* peak emission wavelength.

peak-to-average ratio (*p/a* r). The ratio of the instantaneous peak value, *i.e.*, maximum magnitude, of a signal parameter to its time-averaged value. *Note:* The peak-to-average ratio can be determined for many signal parameters, such as voltage, current, power, frequency, and phase.

peak-to-peak value. The absolute value of the difference between the maximum and the minimum magnitudes of a varying quantity.

peak wavelength. 1. *Synonym* peak emission wavelength. 2. Of an optical bandpass filter, the wavelength that suffers the lowest loss.

peer entity. In layered systems, one of a set of entities that are in the same layer or the equivalent layer of another system.

peer group. In Open Systems Interconnection (OSI)–Architecture, a group of functional units in a given layer of a network in which all the functions performed by the functional units extend throughout the system at the same layer.

peg count. 1. In communication systems, a count that is made of the number of times that an event or condition occurs. 2. In telephone systems, the process that provides counts of the calls of different service classes that occur during intervals of such frequency as to reliability indicate the traffic load. 3. A count of the attempts to seize, or a count of the actual seizures that occur, of various types of telephone trunks, access lines, switches, or other equipment.

pel. In a facsimile system, the smallest discrete scanning line sample containing only monochrome information, *i.e.*, not containing gray-scale information.

penetration. 1. The passage through a partition or wall of an equipment or enclosure by a wire, cable, or other conductive object. 2. [The] unauthorized act of bypassing the security mechanisms of a cryptographic system or AIS. 3. The passage of a radio frequency through a physical barrier, such as a partition, a wall, a building, or earth.

PEP. *Deprecated abbreviation for* peak envelope power. Either *"PX"* or *"pX"* is now preferred. *See* peak envelope power, power.

percentage modulation. 1. In angle modulation, the fraction of a specified reference modulation, expressed in percent. 2. In amplitude modulation, the modulation factor expressed in percent. *Note:* Percentage modulation may also be expressed in dB below 100% modulation.

percent break. In telephone dialing, the ratio, expressed in percent, of the open circuit time to the sum of the open and closed circuit times allotted to a single dial pulse cycle.

performance management. In network management, (a) a set of functions that evaluate and report the behavior of telecommunications equipment and the effectiveness of the network or network element and (b) a set of various subfunctions, such as gathering statistical information, maintaining and examining historical logs, determining system performance under natural and artificial conditions, and altering system modes of operation.

performance measurement period. The period during which performance parameters are measured. *Note:* A performance measurement period is determined by required confidence limits and may vary as a function of the observed parameter values. User time is divided into consecutive performance measurement periods to enable measurement of user information transfer reliability.

performance parameter. A quality, usually quantified by a numerical value, which quality characterizes a particular aspect, capability, or attribute of a system. *Note:* Examples of performance parameters are peg count and mean time between failures.

periapsis. In a satellite orbit, the point that is closest to the gravitational center of the system consisting of the primary body and the satellite. *Note:* In an orbit about the Earth, periapsis is called *perigee.* In an orbit about the Moon, periapsis is called *perilune,* and in an orbit about the Sun, it is called *perihelion.*

perigee. Of a satellite orbiting the Earth, the point in the orbit at which the gravitational centers of the satellite and Earth are closest to one another.

perigee altitude. *See* altitude of the apogee or of the perigee.

periodic antenna. An antenna that has an approximately constant input impedance over a narrow range of frequencies. *Note:* An example of a periodic antenna is a dipole array antenna. *Synonym* resonant antenna.

period (of a satellite). The time elapsing between two consecutive passages of a satellite through a characteristic point on its orbit.

periods processing. The processing of various levels of classified or unclassified information at distinctly different times. *Note:* Under periods processing, the system must be purged of all information from one processing period before transitioning to the next when there are different users with differing authorizations.

peripheral device. *See* peripheral equipment.

peripheral equipment. In a data processing system, any equipment, distinct from the central processing unit, that may provide the system with additional capabilities. *Note:* Such equipment is often offline until needed for a specific purpose and may, in some cases, be shared among several users.

peripheral node. *Synonym* endpoint node.

periscope antenna. An antenna configuration in which the transmitting antenna is oriented to produce a vertical radiation pattern, and a flat or off-axis parabolic reflector, mounted above the transmitting antenna, is used to direct the beam in a horizontal path toward the receiving antenna. *Note:* A periscope antenna facilitates increased terrain clearance without long transmission lines, while permitting the active equipment to be located at or near ground level for ease of maintenance.

permanent bond. A bond not expected to require disassembly for operational or maintenance purposes.

permanent signal (PS). An extended off-hook condition not followed by dialing.

permanent storage. A storage device in which stored data are nonerasable.

permanent virtual circuit (PVC). A virtual circuit used to establish a long-term connection between data terminal equipments (DTE). *Note 1:* In a PVC, the long-term association is identical to the data transfer phase of a virtual call. *Note 2:* Permanent virtual circuits eliminate the need for repeated call set-up and clearing. *Deprecated synonym* nailed-up circuit.

permissible interference. Observed or predicted interference which complies with quantitative interference and sharing criteria contained in these *[Radio] Regulations* or in CCIR Recommendations or in special agreements as provided for in these *Regulations*.

Personal Communications Service. *See* PCS.

personal mobility. In universal personal telecommunications, (a) the ability of a user to access telecommunication services at any UPT terminal on the basis of a personal identifier, and (b) the capability of the network to provide those services in accord with the user's service profile. *Note 1:* Personal mobility involves the network's capability to locate the terminal associated with the user for the purposes of addressing, routing, and charging the user for calls. *Note 2:* "Access" is intended to convey the concepts of both originating and terminating services. *Note 3:* Management of the service profile by the user is not part of personal mobility. The personal mobility aspects of personal communications are based on the UPT number.

personal registration. In universal personal telecommunications, the process of associating a UPT user with a specific terminal.

personal terminal. In personal communications service, a lightweight, small, portable terminal that provides the capability for the user to be either stationary or in motion while accessing and using telecommunication services.

phantom circuit. A third circuit derived from two suitably arranged pairs of wires, called side circuits, with each pair of wires being a circuit in itself and at the same time acting as one conductor of the third circuit. *Note:* The side circuits are coupled to their respective drops by center-tapped transformers, usually called "repeat coils." The center taps are on the line side of the side circuits. Current from the phantom circuit is split evenly by the center taps. This cancels crosstalk from the phantom circuit to the side circuits.

phantom group. Three circuits that are derived from simplexing two physical circuits to form a phantom circuit.

phase. 1. Of a periodic, varying phenomenon, *e.g.*, an electrical signal or electromagnetic wave, any distinguishable instantaneous state of the phenomenon, referred to a fixed reference or another periodic varying phenomenon. *Note 1:* Phase, *i.e., phase time* (frequently abbreviated simply to "phase" in colloquial usage), can be specified or expressed by time of occurrence relative to a specified reference. *Note 2:* The phase of a periodic phenomenon can also be expressed or specified by angular measure, with one period usually encompassing 360° (2π radians). *Note 3:* Phase may be represented (a) in polar coordinates by $M\angle\theta$, where M is the magnitude and θ is the phase angle, and (b) in Cartesian coordinates, *i.e.*, an Argand diagram, as $(a + jb)$, where a is a real component and b is an imaginary component such that $\tan \theta = (b/a)$, where θ is the phase angle, and the magnitude, M, is $(a^2 + b^2)^{½}$ 2. A distinguishable state of a phenomenon. 3. That period of time during which a specified function occurs in a sequential list of functions.

phase angle. Of a periodic wave, the number of suitable units of angular measure between a point on the wave and a reference point. *Note 1:* The reference point may be a point on another periodic wave. The waves may be plotted on a suitable coordinate system, such as a Cartesian plot, with degrees or other angular measure usually plotted on the abscissa and amplitude on the ordinate. Usually, at least one full cycle of each wave is plotted, with 360° (2π radians) encompassing one full cycle. The reference points may be any significant instants on the waves, such as where they cross the abscissa axis. *Note 2:* The use of angular measure to define the relationship between a periodic wave and a reference point is derived from the projection of a rotating vector onto the real axis of an Argand diagram. *Note 3:* The value of the phase angle of a point on the wave is the point on the abscissa that corresponds to the point on the wave. *Note 4:* The phase angle of a vector may be written as $M\angle\theta$, where M is the magnitude of the vector and θ is the phase angle relative to the specified reference.

phase bandwidth. Of a network or device, the width of the continuous frequency range over which the phase-vs.-frequency characteristic does not depart from linearity by more than a stated amount.

phase coherence. The state in which two signals maintain a fixed phase relationship with each other or with a third signal that can serve as a reference for each.

phase coherent. *See* phase coherence.

phase constant. The imaginary part of the axial propagation constant for a particular mode, usually expressed in radians per unit length.

phased array. A group of antennas in which the relative phases of the respective signals feeding the antennas are varied in such a way that the effective radiation pattern of the array is reinforced in a desired direction and suppressed in undesired directions. *Note 1:* The relative amplitudes of—and constructive and destructive interference effects among—the signals radiated by the individual antennas determine the effective radiation pattern of the array. *Note 2:* A phased array may be used to point a fixed radiation pattern, or to scan rapidly in azimuth or elevation.

phase delay. In the transmission of a single-frequency wave from one point to another, the delay of an arbitrary point in the wave that identifies its phase. *Note:* Phase delay may be expressed in any convenient unit, such as seconds, degrees, radians, or wavelengths.

phase departure. 1. A phase deviation from a specified value. 2. An unintentional deviation from the nominal phase value.

phase detector. A circuit or instrument that detects the difference in phase between corresponding points on two signals.

phase deviation. In phase modulation, the maximum difference between the instantaneous phase angle of the modulated wave and the phase angle of the unmodulated carrier. *Note:* For a sinusoidal modulating wave, the phase deviation, expressed in radians, is equal to the modulation index.

phase diagram. A graphic representation of the phase relationships between two or more waveforms. *Note:* A phase diagram may be represented as a vector diagram or as an amplitude-vs.-time diagram.

phase difference. The time interval or phase angle by which one wave leads or lags another. *Synonym* **phase offset.**

phase distortion. Distortion that occurs when (a) the phase-frequency characteristic is not linear over the frequency range of interest, *i.e.*, the phase shift introduced by a circuit or device is not directly proportional to frequency, or (b) the zero-frequency intercept of the phase-frequency characteristic is not 0 or an integral multiple of 2ð radians. *Synonym* phase-frequency distortion.

phase-encoded (PE) recording. Binary recording on magnetic media, such as magnetic disks, tapes, and cards, in which a "1" is represented by a magnetic flux reversal to the polarity of the interblock gap, and a "0" is represented by a magnetic flux reversal to the polarity opposite to that of the interblock gap when recording in the forward direction.

phase equalizer. *See* delay equalizer.

phase flux reversal. In phase-encoded recording, a magnetic flux reversal written at the nominal midpoint between successive "1" bits, or between successive "0" bits, to establish proper polarity.

phase-frequency characteristic. A Cartesian-coordinate plot of phase shift as the dependent variable, versus frequency as the independent variable. *Note:* The phase-frequency characteristic is linear if the phase shift introduced by a circuit or device is the same for all frequencies in the input signal.

phase-frequency distortion. *Synonym* phase distortion.

phase hit. *See* hit.

phase instability. The fluctuation of the phase of a wave, relative to a reference. *Note:* The fluctuation is often from unknown causes.

phase interference fading. The variation in signal amplitude produced by the interaction of two or more signal elements with different relative phases.

phase inversion. Introduction of a phase difference of 180E. *Note:* Phase inversion may occur with a random or periodic, symmetrical or non-symmetrical waveform, although it is usually produced by the inversion of a symmetrical periodic signal, resulting in a change in sign. A symmetrical periodic signal represented by $f(t) = Ae^{j\omega t}$, after phase inversion, becomes $f_1(t) = Ae^{j(\omega t+\pi)}$, where t is time, A is the magnitude of the vector, ω is angular frequency ($\omega = 2\pi f$), where f is the frequency and $\omega \approx 3.1416$ and $e \approx 2.7183$. The algebraic sum of $f(t)$ and $f_1(t)$ will always be zero.

phase jitter. Rapid, repeated phase perturbations that result in the intermittent shortening or lengthening of signal elements. *Note 1:* Phase jitter may be random or cyclic. *Note 2:* The phase departure in phase jitter usually is smaller, but more rapid, than that of phase perturbation. Phase jitter may be expressed in degrees, radians, or seconds. Phase jitter is usually random. However, if cyclic, phase jitter may be expressed in hertz as well as in degrees, radians, or seconds.

phase jump. A sudden phase change in a signal.

phase-locked loop (PLL). An electronic circuit that controls an oscillator so that it maintains a constant phase angle relative to a reference signal. *Note:* Phase-locked loops are widely used in space communications for coherent carrier tracking and threshold extension, bit synchronization, and symbol synchronization.

phase measurement tolerance. The maximum allowable difference between a phase measurement and the actual phase value.

phase modulation (PM). Angle modulation in which the phase angle of a carrier is caused to depart from its reference value by an amount proportional to the instantaneous value of the modulating signal.

phase noise. In an oscillator, rapid, short-term, random fluctuations in the phase of a wave, caused by time-domain instabilities. *Note:* Phase noise, $\mathcal{L}(f)$ in decibels relative to carrier power (dBc) on a 1-Hz bandwidth, is given by $\mathcal{L}(f) = 10\log[0.5(S_\phi(f))]$ where $S_\phi(f)$ is the spectral density of phase fluctuations.

phase nonlinearity. Lack of direct proportionality of phase shift to frequency over the frequency range of interest.

phase offset. *Synonym* phase difference.

phase perturbation. Any shifting (often quite rapid), from whatever cause, in the phase of a signal. *Note 1:* The shifting in phase may appear to be random, cyclic, or both. *Note 2:* The phase departure in phase perturbation usually is larger, but less rapid, than that of phase jitter. *Note 3:* Phase perturbation may be expressed in degrees, with any cyclic component expressed in hertz.

phase quadrature. *See* quadrature.

phase shift. The change in phase of a periodic signal with respect to a reference.

phase-shift keying (PSK). 1. In digital transmission, angle modulation in which the phase of the carrier is discretely varied in relation either to a reference phase or to the phase of the immediately preceding signal element, in accordance with data being transmitted. 2. In a communications system, the representing of characters, such as bits or quaternary digits, by a shift in the phase of an electromagnetic carrier wave with respect to a reference, by an amount corresponding to the symbol being encoded. *Note 1:* For example, when encoding bits, the phase shift could be 0° for encoding a "0," and 180° for encoding a "1," or the phase shift could be −90° for "0" and +90° for a "1," thus making the representations for "0" and "1" a total of 180° apart. *Note 2:* In PSK systems designed so that the carrier can assume only two different phase angles, each change of phase carries one bit of information, *i.e.,* the bit rate equals the modulation rate. If the number of recognizable phase angles is increased to 4, then 2 bits of information can be encoded into each signal element; likewise, 8 phase angles can encode 3 bits in each signal element. *Synonyms* biphase modulation, phase-shift signaling.

phase-shift signaling. *Synonym* phase-shift keying.

phase term. In the propagation of an electromagnetic wave in a uniform waveguide, such as an optical fiber or metal waveguide, the parameter that indicates the phase change per unit distance of the wave at any point along the waveguide.

phase velocity. The velocity of propagation of a uniform plane wave, given by (a) the product of the wavelength and the frequency divided by (b) the refractive index of the medium in which the wave is propagating. *Note 1:* In free space, the refractive index may be considered as unity. *Note 2:* In free space, the group velocity and the phase velocity are equal.

phasing. In facsimile transmission and reception, the process by which the start of the scanning line or lines is made to correspond to one edge of the object being scanned. *Note:* If there is no correspondence between the object being scanned and the scanning line or lines, distortion, often in the form of a split image, will occur in the received image.

phon. In acoustics, a unit of subjective loudness level equal to the sound pressure level in dB compared to that of an equally loud standard sound. *Note:* The accepted standard is a 1-kHz pure sine-wave tone or narrowband noise centered at 1 kHz.

phone. 1. *Abbreviation for* telephone, telephony. 2. *Colloquially,* the voice-operation mode in radio communications.

phonetic alphabet. A list of standard words used to identify letters in a message transmitted by radio or telephone. The following are the currently authorized words for each letter in the alphabet: Alpha, Bravo, Charlie, Delta, Echo, Foxtrot, Golf, Hotel, India, Juliet, Kilo, Lima, Mike, November, Oscar, Papa, Quebec, Romeo, Sierra, Tango, Uniform, Victor, Whiskey, X-ray, Yankee, Zulu.

phonon. A quantum of acoustic energy, the level of which is a function of the frequency of the acoustic wave. *Note:* Phonons in acoustics are analogous to photons in electromagnetics. The energy of a phonon is usually less than 0.1 eV (electron-volt) and thus is one or two orders of magnitude less than that of a photon. When photons and phonons interact in semiconductors used in communications systems, undesirable system behavior can occur.

phonon absorption. Absorption of light energy by its conversion to vibrational energy. *Note:* Phonon absorption determines the fundamental, *i.e.,* quantum limit of attenuation, *i.e.,* minimum attenuation, in silica-based glasses in the far infrared region.

photoconductive effect. In certain materials, the phenomenon that results in photoconductivity.

photoconductivity. In certain materials, the increase in electrical conductivity that results from increases in the number of free carriers generated when photons are absorbed. *Note:* The photons must have quantum energy sufficient to overcome the band-gap in the material in question.

photocurrent. The current that flows through a photosensitive device, such as a photodiode, as the result of exposure to radiant power. *Note 1:* The photocurrent may occur as a result of the photoelectric, photoemissive, or photovoltaic effect. *Note 2:* The photocurrent may be enhanced by internal gain caused by interaction among ions and photons under the influence of applied fields, such as occurs in an avalanche photodiode (APD).

photodetector (PD). A transducer capable of accepting an optical signal and producing an electrical signal containing the same information as in the optical signal. *Note:* The two main types of semiconductor photodetectors are the photodiode (PD) and the avalanche photodiode (APD).

photodiode. A semiconductor diode that produces, as a result of the absorption of photons, (a) a photovoltage or (b) free carriers that support the conduction of photocurrent. *Note:* Photodiodes are used for the detection of optical communication signals and for the conversion of optical power to electrical power.

photoelectric effect. In certain materials, the changes in the electrical characteristics caused by photon absorption.

photon. A discrete packet, *i.e.*, quantum, of electromagnetic energy. *Note:* The energy of a photon is hv, where h is Planck's constant and v is the frequency of the electromagnetic wave.

photon noise. In an optical communication link, noise attributable to the statistical nature of optical quanta. *See* quantum noise.

photosensitive recording. Facsimile recording by the exposure of a photosensitive surface to a signal-controlled light beam or spot.

photovoltaic effect. The production, as a result of the absorption of photons, of a voltage difference across a pn junction. *Note:* The voltage difference is caused by the internal drift of holes and electrons.

physical frame. *See* frame.

Physical Layer. *See* Open Systems Interconnection—Reference Model.

physical optics. The branch of optics that treats light propagation as a wave phenomenon rather than a ray phenomenon, as in geometric optics.

physical security. *See* communications security.

physical signaling sublayer (PLS). In a local area network (LAN) or a metropolitan area network (MAN) using open systems interconnection (OSI) architecture, the portion of the physical layer that (a) interfaces with the medium access control sublayer, (b) performs character encoding, transmission, reception, and decoding, and (c) performs optional isolation functions.

physical topology. The physical configuration, *i.e.*, interconnection, of network elements, *e.g.*, cable paths, switches, concentrators. *Note:* Physical topology is in contrast to logical topology. For example, a logical loop may consist of a physical star configuration, or a physical loop.

picowatt. *See* pW.

picowatt, psophometrically weighted. *See* noise weighting.

picture. *Synonym* image.

picture black. In TV and facsimile, pertaining to the signal or signal level that corresponds to the darkest part, *i.e.*, the spot with the lowest luminance or reflectivity, of the object being scanned.

picture element. *See* pel, pixel.

picture frequency. In analog facsimile systems, a baseband frequency generated by scanning an object. *Note:* Picture frequencies do not include frequencies that are present in a modulated carrier.

picture white. In TV and facsimile, pertaining to the signal level that corresponds to the brightest part, *i.e.*, the spot with the highest luminance or reflectivity, of the object being scanned.

piecewise linear encoding. *See* segmented encoding law.

piecewise linear encoding law. *Synonym* segmented encoding law.

pigtail. **1**. A short length of optical fiber that is permanently affixed to an active device, *e.g,* LED or laser diode, and is used to couple the device, using a splice or connector, to a longer fiber. **2**. A short length of single-fiber cable, usually tight-buffered, that has an optical connector on one end and a length of exposed fiber at the other end. *Note:* The exposed fiber of the pigtail is then spliced to one fiber of a multifiber trunk, *i.e.,* arterial, cable, to enable the multifiber cable to be "broken out" into individual single-fiber cables that may be connected to a patch panel or an input or output port of an optical receiver or transmitter. 3. A short length of electrical conductor permanently affixed to a component, used to connect the component to another conductor.

pilot. A signal, usually a single frequency, transmitted over a communications system for supervisory, control, equalization, continuity, synchronization, or reference purposes. *Note:* Sometimes it is necessary to employ several independent pilot frequencies. Most radio relay systems use radio or continuity pilots of their own but transmit also the pilot frequencies belonging to the carrier frequency multiplex system.

pilot frequency. *See* synchronizing pilot.

pilot-make-busy (PMB) circuit. A circuit arrangement by which trunks provided over a carrier system are made busy to the switching equipment in the event of carrier system failure, or during a fade of the radio system.

pilot tone. *See* pilot.

PIN diode. *Acronym for* positive-intrinsic-negative diode. A photodiode with a large, neutrally doped intrinsic region sandwiched between p-doped and n-doped semiconducting regions. *Note:* A PIN diode exhibits an increase in its electrical conductivity as a function of the intensity, wavelength, and modulation rate of the incident radiation. *Synonym* PIN photodiode.

ping. *Abbreviation for* packet Internet groper. In TCP/IP, a protocol function that tests the ability of a computer to communicate with a remote computer by sending a query and receiving a confirmation response.

pink noise. In acoustics, noise in which there is equal power per octave.

PIN photodiode. *Synonym* PIN diode.

piston. In a hollow metallic waveguide, a longitudinally movable metallic plane surface that reflects essentially all the incident energy. *Note:* A piston is used for tuning, *e.g.*, fine-tuning a resonant cavity. *Synonym* plunger.

pitch. *Synonym* lay length.

pixel. In a raster-scanned imaging system, the smallest discrete scanning line sample that can contain gray scale information.

PLA. *Abbreviation for* programmable logic array.

plain text. Unencrypted information. *Note:* Plain text includes voice. *Synonym* clear text.

planar array. An antenna in which all of the elements, both active and parasitic, are in one plane. *Note 1:* A planar array provides a large aperture and may be used for directional beam control by varying the relative phase of each element. *Note 2:* A planar array may be used with a reflecting screen behind the active plane.

planar waveguide. *Synonym* slab-dielectric waveguide.

Planck's constant. The constant of proportionality, represented by the symbol h, that relates the energy E of a photon with the frequency v of the associated wave through the relation E = hv, where h = 6.626×10^{-34} joule•second.

Planck's law. The fundamental law of quantum theory that describes the essential concept of the quanta of electromagnetic energy. *Note 1:* Planck's law states that the quantum of energy, E, associated with an electromagnetic field is given by E = hv, where h is Planck's constant and v is the frequency of the electromagnetic radiation. *Note 2:* Planck's constant is usually given in joule•seconds and the frequency in hertz. Thus, the quantum of energy is usually given in joules. *Note 3:* The product of energy and time is sometimes referred to as the elementary quantum of action. Hence, h is sometimes referred to as the elementary quantum of action.

plane polarization. *Synonym* linear polarization.

plane wave. 1. A wave whose surfaces of constant phase are infinite parallel planes normal to the direction of propagation. 2. An electromagnetic wave that predominates in the far-field region of an antenna, and has a wavefront that is essentially in a plane. *Note:* In free space, the characteristic impedance of a plane wave is 377 Ω.

plant. All the facilities and equipment used to provide telecommunications services. *Note:* Plant is usually characterized as *outside plant* or *inside plant*. Outside plant, for example, includes all poles, repeaters and unoccupied buildings housing them, ducts, and cables–including the "inside" portion of interfacility cables outward from the main distributing frame (MDF) in a central office or switching center. Inside plant includes the MDF and all equipment and facilities within a central office or switching center.

plastic-clad silica (PCS) fiber. An optical fiber that has a silica-based core and a plastic cladding. *Note 1:* The cladding of a PCS fiber should not be confused with the polymer overcoat of a conventional all-glass fiber. *Note 2:* PCS fibers in general have significantly lower performance characteristics, *i.e.,* higher transmission losses and lower bandwidths, than all-glass fibers. *Synonym* polymer-clad silica fiber.

plenum. In a building, an enclosure, created by building components such as a suspended ceiling or false floor, and used for the movement of environmental air. *Note 1:* A plenum may be used to contain communications and power cables, *e.g.,* to reach equipment installed in open office or laboratory space. *Note 2:* Cables installed in plenums must meet applicable environmental and fire protection regulations. This may mean enclosing them in suitable ducts or using cables having jackets and other components made of materials that are resistant to open flame and are non-toxic at high temperatures.

plesiochronous. That relationship between two signals such that their corresponding significant instants occur at nominally the same rate, any variations being constrained within a specified limit. *Note:* There is no limit to the phase difference that can accumulate between corresponding significant instants over a long period of time.

PL/I. A programming language that is designed for use in a wide range of commercial and scientific computer applications.

PLL. *Abbreviation for* phase-locked loop.

plotter. An output unit that presents data in the form of a two-dimensional graphic representation.

PLS. *Abbreviation for* physical signaling sublayer.

plunger. *Synonym* piston.

PM. *Abbreviation for* phase modulation, preventive maintenance.

PMB. *Abbreviation for* pilot-make-busy. *See* pilot-make-busy circuit.

Pockels cell. An electro-optic device in which birefringence is modified under the influence of an applied voltage. *Note:* A Pockels cell may be used as an intensity modulator at optical wavelengths.

POI. *Abbreviation for* point of interface.

pointer. 1. A function indicator that (a) is under the direct control of a computer operator, and (b) is used to indicate displayed information, to highlight data, to identify areas of interest, to serve as a graphic display cursor, and/or to select icons. 2. In computer graphics, a manually operated functional unit used to specify an addressable point. 3. In computer programming, an identifier that indicates the location of a data item.

point of interface (POI). In a telecommunications system, the physical interface between the local access and transport area (LATA) access and inter-LATA functions. *Note:* The POI is used to establish the technical interface, the test points, and the points of operational responsibility. *Synonym* interface point.

point of presence (POP). A physical layer within a local access and transport area (LATA) at which an inter-LATA carrier establishes itself for the purpose of obtaining LATA access and to which the local exchange carrier provides access services.

point of train. In infrared transmission systems, a steady infrared light that is used (a) to assist the transmitter in locating a receiving station and (b) for keeping the transmitted light pointed in the proper direction for satisfactory reception.

point source. A source of electromagnetic radiation such that (a) the source is so distant from a point of observation or measurement of the radiation that the wavefront of the radiation is a planar rather than a curved surface, regardless of the shape of the source, (b) the size or shape of the source has no influence on the shape of the wavefront at the point of observation or measurement, and (c) the source need not necessarily radiate with equal radiance in all directions.

point-to-point link. A dedicated data link that connects only two stations.

point-to-point transmission. Communications between two designated stations only.

Poisson distribution. A mathematical statement of the probability that exactly *k* discrete events will take place during an interval of length *t*, expressed by

$$P(k, t) = \frac{(\lambda t)^{t} e^{-\lambda t}}{k!},$$

where *k* is a non-negative integer, e is the base of the natural logarithms (e2.71828), λ is the constant rate that the events occur, and *λt* is the expected number of events occurring during an interval of length *t*.

polar direct-current telegraph transmission. A form of binary telegraph transmission in which positive and negative direct currents denote the significant conditions. *Synonym* double-current transmission.

polarential telegraph system. A direct-current telegraph system employing polar transmission in one direction and a form of differential duplex transmission in the other. *Note:* Two types of polarential systems, known as types A and B, are in use. In half-duplex operation of a type A polarential system, the direct-current balance is independent of line resistance. In half-duplex operation of a type B polarential system, the direct current is substantially independent of the line leakage. Type A is better for cable loops where leakage is negligible but resistance varies with temperature. Type B is better for open wire where variable line leakage is frequent.

polarization. Of an electromagnetic wave, the property that describes the orientation, *i.e.*, time-varying direction and amplitude, of the electric field vector. *Note 1:* States of polarization are described in terms of the figures traced as a function of time by the projection of the extremity of a representation of the electric vector onto a fixed plane in space, which plane is perpendicular to the direction of propagation. In general, the figure, *i.e.*, polarization, is elliptical and is traced in a clockwise or counterclockwise sense, as viewed in the direction of propagation. If the major and minor axes of the ellipse are equal, the polarization is said to be *circular*. If the minor axis of the ellipse is zero, the polarization is said to be *linear*. Rotation of the electric vector in a clockwise sense is designated *right-hand polarization*, and rotation in a counterclockwise sense is designated *left-hand polarization*. *Note 2:* Mathematically, an elliptically polarized wave may be described as the vector sum of two waves of equal wavelength but unequal amplitude, and in quadrature (having their respective electric vectors at right angles and ð/2 radians out of phase).

polarization diversity. Diversity transmission and reception wherein the same information signal is transmitted and received simultaneously on orthogonally polarized waves with fade-independent propagation characteristics

polarization-maintaining (PM) optical fiber. An optical fiber in which the polarization planes of lightwaves launched into the fiber are maintained during propagation with little or no cross-coupling of optical power between the polarization modes. [2196] *Note 1:* Cross sections of polarization-maintaining optical fibers range from elliptical to rectangular. *Note 2:* Polarization-maintaining optical fibers are used in special applications, such as in fiber optic sensing and interferometry. *Synonym* polarization-preserving (PP) optical fiber.

polarization-preserving (PP) optical fiber. *Synonym* polarization-maintaining optical fiber.

polar operation. A telegraph system in which marking signals are formed by current or voltage pulses of one polarity and spacing signals by current or voltage pulses of equal magnitude but opposite polarity (bipolar signal).

polar orbit. An orbit for which the angle of inclination is 90°. *Note:* A satellite in polar orbit will pass over both the north and south geographic poles once per orbit.

polar relay. A dc relay in which the direction of movement of the armature depends on the direction of the current flow.

polling. 1. Network control in which the control station invites tributary stations to transmit in the sequence specified by the control station. 2. In point-to-point or multipoint communication, the process whereby stations are invited one at a time to transmit. 3. Sequential interrogation of devices for various purposes, such as avoiding contention, determining operational status, or determining readiness to send or receive data. 4. In automated HF radio systems, a technique for measuring and reporting channel quality.

polymer-clad silica fiber. *Synonym* plastic-clad silica fiber.

POP. *Acronym for* point of presence.

port. 1. Of a device or network, a point of access where signals may be inserted or extracted, or where the device or network variables may be observed or measured. 2. In a communications network, a point at which signals can enter or leave the network en route to or from another network.

portability. 1. The ability to transfer data from one system to another without being required to recreate or reenter data descriptions or to modify significantly the application being transported. 2. The ability of software or of a system to run on more than one type or size of computer under more than one operating system. *See* POSIX. 3. Of equipment, the quality of being able to function normally while being conveyed.

portable station. 1. A station capable of being carried by one or more persons. *Note:* A portable station usually has a self-contained power source and can be operated while being carried. 2. A station designed to be carried by a person and capable of transmitting and/or receiving while in motion or during brief halts at unspecified locations.

port operations service. A maritime mobile service in or near a port, between coast stations and ship stations, or between ship stations, in which messages are restricted to those relating to the operational handling, the movement and the safety of ships and, in emergency, to the safety of persons. Messages which are of a public correspondence nature shall be excluded from this service.

portrait mode. 1. In facsimile, the mode of scanning lines across the shorter dimension of a rectangular original. *Note:* CCITT Group 1, 2, and 3 facsimile machines use portrait mode. 2. In computer graphics, the orientation of an image in which the shorter dimension of the image is horizontal. 3. An orientation of printed text on a page such that the lines of text are perpendicular to the long dimension of the page.

port station. A coast station in the port operations service.

positioned channel. In integrated services digital networks (ISDN), a channel that occupies dedicated bit positions in the framed data stream. *Note:* Examples of positioned channels are the B, H, and D channels.

positioned interface structure. Within a framed interface, a structure in which positioned channels provide all services and signaling.

positioning time. *Synonym* seek time.

positive feedback. *Synonym* regeneration (def. #1).

positive justification. *Synonym* bit stuffing.

POSIX. *Acronym for* portable operating system interface for computer environments. A Federal Information Processing Standard Publication (FIPS PUB 151-1) for a vendor-independent interface between an operating system and an application program, including operating system interfaces and source code functions. *Note:* IEEE Standard 1003.1-1988 was adopted by reference and published as FIPS PUB 151-1.

postalize. In communications, to structure rates or prices so that they are not distance sensitive, but depend on other factors, such as call duration, type of service, and time of day.

post-detection combiner. *Synonym* maximal-ratio combiner.

post-development review. *Synonym* system follow-up.

post-implementation review. *Synonym* system follow-up.

post-production processing. In broadband ISDN (B-ISDN), applications, the processing of audio and video information after contribution and prior to final use.

power. 1. The rate of transfer or absorption of energy per unit time in a system. 2. Whenever the power of a radio transmitter *etc.* is referred to, it shall be expressed in one of the following forms, according to the class of emission, using the arbitrary symbols indicated:

- peak envelope power (PX or pX);
- mean power (PY or pY);
- carrier power (PZ or pZ).

For different classes of emission, the relationships between peak envelope power, mean power and carrier power, under the conditions of normal operation and of no modulation, are contained in CCIR Recommendations which may be used as a guide. For use in formulae, the symbol p denotes power expressed in watts and the symbol P denotes power expressed in decibels relative to a reference level.

power budget. The allocation, within a system, of available electrical power, among the various functions that need to be performed. *Note:* An example of a power budget in a communications satellite is the allocation of available power among various functions, such as maintaining satellite orientation, maintaining orbital control, performing signal reception, and performing signal retransmission. *Synonym* system budget.

power circuit breaker (PCB). 1. The primary switch used to apply and remove power from equipment. 2. A circuit breaker used on ac circuits rated in excess of 1500V.

power density. *Deprecated synonym for* irradiance.

power factor. In alternating-current power transmission and distribution, the cosine of the phase angle between the voltage and current. *Note 1:* When the load is inductive, *e.g.,* an induction motor, the current lags the applied voltage, and the power factor is said to be a *lagging* power factor. When the load is capacitive, *e.g.,* a synchronous motor or a capacitive network, the current leads the applied voltage, and the power factor is said to be a *leading* power factor. *Note 2:* Power factors other than unity have deleterious effects on power transmission systems, including excessive transmission losses and reduced system capacity. Power companies therefore require customers, especially those with large loads, to maintain, within specified limits, the power factors of their respective loads or be subject to additional charges.

power failure transfer. 1. The switching of primary utilities to their secondary backup whenever the primary source operates outside its design parameters. 2. In telephony, a function, which, when activated in the event of a commercial power failure or a low-voltage battery condition at a subscriber location, supplies power to predesigned subscriber equipment via the central office trunk. *Note:* Power-failure transfer is an emergency mode of operation in which one and only one instrument may be powered from each trunk line from the subscriber location to the central office.

power gain of an antenna. *Synonym* antenna gain.

power-law index profile. For optical fibers, a class of graded-index profiles characterized by

$$n(r) = \begin{cases} n_1\sqrt{1 - 2\Delta\left(\dfrac{r}{\alpha}\right)^g}, & r \leq \alpha \\ n_1\sqrt{1 - 2\Delta}, & r \geq \alpha \end{cases}$$

$$\text{where } \Delta = \frac{n_1^2 - n_2^2}{2n_1^2},$$

where $n(r)$ is the nominal refractive index as a function of distance from the fiber axis, n_1 is the nominal refractive index on axis, n_2 is the refractive index of the homogeneous cladding ($n(r) = n_2$ when $r \geq \alpha$), α is the core radius, and g is a parameter that defines the shape of the profile. *Note 1:* α is often used in place of g. Hence, this is sometimes called an alpha profile. *Note 2:* For this class of profiles, multimode distortion is smallest when g takes a particular value depending on the material used. For most materials, this optimum value is approximately 2. When g increases without limit, the profile tends to a step-index profile.

power margin. The difference between available signal power and the minimum signal power needed to overcome system losses and still satisfy the minimum input requirements of the receiver for a given performance level. *Note:* System power margin reflects the excess signal level, present at the input of the receiver, that is available to compensate for (a) the effects of component aging in the transmitter, receiver, or physical transmission medium, and (b) a deterioration in propagation conditions. *Synonym* system power margin.

Poynting vector. The vector obtained in the direction of a right-hand screw from the cross-product (vector product) of the electric field vector rotated into the magnetic field vector of an electromagnetic wave. *Note:* The Poynting vector, with transmission media parameters and constants, gives the irradiance and direction of propagation of the electromagnetic wave. Mathematically: $P = E \times H$.

PPM. *Abbreviation for* pulse-position modulation.

pre-arbitrated slot. A slot dedicated by the head-of-bus function for transferring isochronous service octets.

preassignment access plan. In satellite communications system operations, a fixed communication channel access plan, as opposed to a demand assignment access plan in which allocation of accesses or the number of channels per access is varied in accordance with the demand.

precedence. In communications, a designation assigned to a message by the originator to indicate to communications personnel the relative order of handling and to the addressee the order in which the message is to be noted. *Note:* The descending order of precedence for military messages is FLASH, IMMEDIATE, PRIORITY, and ROUTINE.

precipitation attenuation. The loss of energy by an electromagnetic wave because of scattering, refraction, and/or absorption during its passage through a volume of the atmosphere containing precipitation such as rain, snow, hail, or sleet.

precipitation static (p-static). Radio interference caused by the impact of charged particles against an antenna. *Note:* Precipitation static may occur in a receiver during certain weather conditions, such as snowstorms, hailstorms, rainstorms, dust storms, or combinations thereof.

precise frequency. A frequency that is maintained to the known accuracy of an accepted reference frequency standard. *Note:* Current uncertainty among international standards is approximately 1 part in 10^{14} as of 1995.

precise time. A time mark that is accurately known with respect to an accepted reference time standard. *Note:* Current uncertainty among international standards is approximately 1 part in 10^{14} as of 1995.

precise time and time interval (PTTI). The discipline that addresses precise timekeeping and time information transfer.

precision. 1. The degree of mutual agreement among a series of individual measurements, values, or results; often, but not necessarily, expressed by the standard deviation. 2. With respect to a set of independent devices of the same design, the ability of these devices to produce the same value or result, given the same input conditions and operating in the same environment. 3. With respect to a single device, put into operation repeatedly without adjustments, the ability to produce the same value or result, given the same input conditions and operating in the same environment. *Synonym (for defs. 1, 2, and 3)* reproducibility. 4. In computer science, a measure of the ability to distinguish between nearly equal values. 5. The degree of discrimination with which a quantity is stated; for example, a three-digit numeral to the base 10 discriminates among 1000 possibilities.

precision-sleeve splicing. Optical fiber splicing that uses a capillary tube, of suitable material, to align the mating fibers. *Note:* The capillary tube has an inside diameter slightly larger than the cladding diameter of the two optical fibers to be spliced. The fibers are inserted, one from either end, to form a butt joint. The capillary tube may contain an index-matching gel, or the fibers may be secured with an adhesive having a refractive index that approximates that of the fibers.

precombining. The combining of multiplexed signals prior to the modulation of the carrier. *Synonym* premodulation combining.

predetection. Referring to that portion of the circuitry of a receiver which, with respect to the signal being processed, is chronologically prior to the detection. *Note:* Predetection signals contain the carrier signal and all modulation, and are basically at radio frequencies.

predetection combining. *Synonym* maximal-ratio combiner.

preemphasis. A system process designed to increase, within a band of frequencies, the magnitude of some (usually higher) frequencies with respect to the magnitude of other (usually lower) frequencies, in order to improve the overall signal-to-noise ratio by minimizing the adverse effects of such phenomena as attenuation differences, or saturation of recording media, in subsequent parts of the system. *Note:* Preemphasis has applications, for example, in audio recording and FM transmission.

preemphasis improvement. In FM broadcasting, the improvement in the signal-to-noise ratio of the high-frequency portion of the baseband, *i.e.,* modulating, signal, which improvement results from passing the modulating signal through a preemphasis network. *Note:* Preemphasis increases the magnitude of the higher signal frequencies, thereby improving the signal-to-noise ratio. At the output of the discriminator in the FM receiver, a deemphasis network restores the original signal power distribution.

preemphasis network. A network inserted in a system in order to increase the magnitude of one range of frequencies with respect to another. *Note:* Preemphasis is usually employed in FM or phase modulation transmitters to equalize the modulating signal drive power in terms of deviation ratio. The receiver demodulation process includes a reciprocal network, called a deemphasis network, to restore the original signal power distribution.

preempting call. *See* multilevel precedence and preemption.

preemption. The seizure, usually automatic, of military system facilities that are being used to serve a lower precedence call in order to serve immediately a higher precedence call.

preemption tone. In military telephone systems, a distinctive tone that is used to indicate to connected users, *i.e.*, subscribers, that their call has been preempted by a call of higher precedence. *Note:* An example of preemption tone is a distinctive, steady, high-pitch tone transmitted for three seconds or until the preempted user hangs up.

prefix-free code. *Synonym* comma-free code.

pregroup combining. In communications systems, assembling a number of narrowband channels, such as 4-kHz-wide telephone channels, into a specified frequency band such that, after pregroup translation, they may be formed with other pregroups into a standard group, such as a CCITT basic group, by frequency-division multiplexing.

pregroup translation. In communications systems, the process of transposing, in frequency, a pregroup of channels, such as telephone or data channels, in such a manner that they may be formed into a standard group, such as a CCITT basic group, by frequency-division multiplexing.

preliminary call. In radio transmission, a call that (a) includes at least the identification of the calling station and the called station, (b) is designed to establish communications with a particular station, and (c) usually includes a request to the called station to reply, although the request may be implied by the recitation of the call signs. *Note:* A preliminary call may be made on a frequency dedicated to that purpose only, and the rest of the communications session take place on a different frequency or frequencies.

premises wiring. *See* on-premises wiring.

premodulation combining. *Synonym* precombining.

Presentation Layer. *See* Open Systems Interconnection—Reference Model.

preset conference. A service feature that permits the automatic connection of a fixed group of users, or a closed user group with outgoing access, by keying a single directory number.

preset jammer. A jammer in which the frequency of the jamming transmitter is fixed before the transmitter is placed in operation. *Note:* Preset jammers are most useful in airborne jamming operations where weight and space requirements may prohibit the use of operators or elaborate control equipment in flight. Preset jammers are usually used in barrage-jamming over a wide band, usually in overlapping series of frequency bands.

press-to-talk operation. *Synonym* push-to-talk operation.

press-to-type operation. *Synonym* push-to-type operation.

preventive maintenance (PM). 1. The care and servicing by personnel for the purpose of maintaining equipment and facilities in satisfactory operating condition by providing for systematic inspection, detection, and correction of incipient failures either before they occur or before they develop into major defects. 2. Maintenance, including tests, measurements, adjustments, and parts replacement, performed specifically to prevent faults from occurring.

PRF. *Abbreviation for* pulse repetition frequency.

PRI. *Abbreviation for* primary rate interface.

primary channel. 1. The channel that is designated as a prime transmission channel and is used as the first choice in restoring priority circuits. 2. In a communications network, the channel that has the highest data rate of all the channels sharing a common interface. *Note:* A primary channel may support the transfer of information in one direction only, either direction alternately, or both directions simultaneously.

primary coating. The plastic overcoat in intimate contact with the cladding of an optical fiber, applied during the manufacturing process. *Note 1:* The primary coating typically has an outside diameter of approximately 250 to 750 μm, and serves to protect the fiber from mechanical damage and chemical attack. It also enhances optical fiber properties by stripping off cladding modes, and in the case where multiple fibers are used inside a single buffer tube, it suppresses cross-coupling of optical signals from one fiber to another. *Note 2:* The primary coating should not be confused with a tight buffer, or the plastic cladding of a plastic-clad-silica (PCS) fiber. *Note 3:* The primary coating, which typically consists of many layers, may be color-coded to distinguish fibers from one another, *e.g.,* in a buffer tube containing multiple fibers. *Synonyms* primary polymer coating, primary polymer overcoat.

primary distribution system. A system of alternating current distribution for supplying the primaries of distribution transformers from the generating station or substation distribution buses.

primary frequency. 1. A frequency that is assigned for usual use on a particular circuit. 2. The first-choice frequency that is assigned to a fixed or mobile station for radiotelephone communications.

primary frequency standard. A frequency source that meets national standards for accuracy and operates without the need for calibration against an external standard. *Note:* Examples of primary frequency standards are hydrogen masers and cesium beam frequency standards.

primary group. *See* group.

primary polymer coating. *Synonym* primary coating.

primary polymer overcoat. *Synonym* primary coating.

primary power. The source of electrical power that usually supplies the station main bus. *Note 1:* The primary power source may be a Government-owned generating plant or a public utility power system. *Note 2:* A Class A primary power source assures, to a high degree of reliability, a continuous supply of ac electrical power.

primary radar. A radiodetermination system based on the comparison of reference signals with radio signals reflected from the position to be determined.

primary radiation. Radiation that is incident upon a material and produces secondary emission from the material.

primary rate interface (PRI). An integrated services digital network (ISDN) interface standard (a) that is designated in North America as having a 23B+D channels, (b) in which all circuit-switched B channels operate at 64 kb/s, and (c) in which the D channel also operates at 64 kb/s. *Note:* The PRI combination of channels results in a digital signal 1 (T1) interface at the network boundary.

primary route. The predetermined path of a message from its source, *i.e.*, sending or originating station, to a message sink, *i.e.*, receiving, addressee, or destination station. *Note 1:* In telephone switchboard operations, the primary route is the route that is attempted first by the operators or equipment when completing a call. *Note 2:* Alternate routing is based on network traffic conditions and supervisory policy.

primary service area. The service area of a broadcast station in which the groundwave is not subject to objectionable interference or objectionable fading.

primary station. In a data communication network, the station responsible for unbalanced control of a data link. *Note:* The primary station generates commands and interprets responses, and is responsible for initialization of data and control information interchange, organization and control of data flow, retransmission control, and all recovery functions at the link level.

primary substation. Equipment that switches or modifies voltage, frequency, or other characteristics of primary power.

primary time standard. A time standard that does not require calibration against another time standard. *Note 1:* Examples of primary time, (*i.e.,* frequency standards) are cesium standards and hydrogen masers. *Note 2:* The international second is based on the microwave frequency (9,192,631,770 Hz) associated with the atomic resonance of the hyperfine ground-state levels of the cesium-133 atom in a magnetically neutral environment. Realizable cesium frequency standards use a strong electromagnet to deliberately introduce a magnetic field which overwhelms that of the Earth. The presence of this strong magnetic field introduces a slight, but known, increase in the atomic resonance frequency. However, very small variations in the calibration of the electric current in the electromagnet introduce minuscule frequency variations among different cesium oscillators.

principal clock. Of a set of redundant clocks, the clock that is selected for normal use. *Note 1:* The principal clock may be selected because of a property, *e.g.*, superior accuracy, that makes it a unique member of the set. *Note 2:* The term *"principal clock"* should not be confused with, or used as a synonym for, the term *"primary frequency standard."*

print-through. A transfer of magnetically recorded data from one part of a data medium to another part of the data medium when these parts are brought into physical contact. *Note:* Print-through is most commonly observed in magnetic tape. Print-through may be avoided by rewinding at timely intervals.

priority. 1. Priority, unless specifically qualified, is the right to occupy a specific frequency for authorized uses, free of harmful interference from stations of other agencies. 2. *Synonym* **priority level.** 3. In DOD record communications systems, one of the four levels of precedence used to establish the time frame for handling a given message. 4. In DOD voice communications systems, one of the levels of precedence assigned to a subscriber telephone for the purpose of preemption of telephone services.

priority level. In the Telecommunications Service Priority system, the level that may be assigned to an NS/EP telecommunications service, which level specifies the order in which provisioning or restoration of the service is to occur relative to other NS/EP or non-NS/EP telecommunication services. *Note:* Priority levels authorized are designated (highest to lowest) "E," "1," "2," "3," "4," and "5" for provisioning and "1," "2," "3," "4," and "5" for restoration. *Synonym* priority.

priority level assignment. The priority level(s) designated for the provisioning or restoration of a particular NS/EP telecommunications service.

priority message. A category of precedence reserved for messages that require expeditious action by the addressee(s) and/or furnish essential information for the conduct of operations in progress when routine precedence will not suffice.

privacy. 1. In a communications system or network, the protection given to information to conceal it from persons having access to the system or network. *Synonym* segregation. 2. In a communications system, protection given to unclassified information, such as radio transmissions of law enforcement personnel, that requires safeguarding from unauthorized persons. 3. In a communications system, the protection given to prevent unauthorized disclosure of the information in the system. *Note 1:* The required protection may be accomplished by various means, such as by communications security measures and by directives to operating personnel. *Note 2:* The limited protection given certain voice and data transmissions by commercial crypto-equipment is sufficient to deter a casual listener, but cannot withstand a competent cryptanalytic attack.

private automatic branch exchange (PABX). *See* PBX.

private automatic exchange (PAX). *See* PBX.

private branch exchange (PBX). *See* PBX.

private exchange (PX). A private telecommunication switch that usually includes access to the public switched network.

private line. In the telephone industry usage, a service that involves dedicated circuits, private switching arrangements, and/or predefined transmission paths, whether virtual or physical, which provide communications between specific locations. *Note:* Among subscribers to the public switched telephone network(s), the term *"private line"* is often used to mean a one-party switched access line.

private NS/EP telecommunications services. Non-common-carrier telecommunications services, including private line, virtual private line, and private switched network services.

procedure-oriented language. A problem-oriented computer programming language that facilitates expressing a procedure in the form of explicit algorithms. *Note:* Examples of procedure-oriented languages are Fortran, ALGOL, COBOL, and PL/I.

proceed-to-select. In communications systems operation, pertaining to a signal or event in the call-access phase of a data call, which signal or event confirms the reception of a call-request signal and advises the calling data terminal equipment to proceed with the transmission of the selection signals. *Note:* Examples of proceed-to-select pertain to a dial tone in a telephone system.

proceed-to-select signal. In a communications system, a signal that indicates that the system is ready to receive a selection signal. *Note:* An example of a proceed-to-select signal is a dial tone.

process computer system. A computer system, with a process interface system, that monitors or controls a technical process.

process control. Automatic control of a process, in which a computer system is used to regulate the usually continuous operations or processes.

process control equipment. Equipment that measures the variables of a technical process, directs the process according to control signals from the process computer system, and provides appropriate signal transformation. *Note:* Examples of process control equipment include actuators, sensors, and transducers.

process control system. A system consisting of a computer, process control equipment, and possibly a process interface system. *Note:* The process interface system may be part of a special-purpose computer.

process gain. In a spread-spectrum communications system, the signal gain, signal-to-noise ratio, signal shape, or other signal improvement obtained by coherent band spreading, remapping, and reconstitution of the desired signal.

processing unit. A functional unit that consists of one or more processors and their internal storage.

process interface system. A functional unit that adapts process control equipment to the computer system in a process computer system.

processor. In a computer, a functional unit that interprets and executes instructions. *Note:* A processor consists of at least an instruction control unit and an arithmetic unit.

procurement. In the Federal Government, the process of obtaining services, supplies, and equipment in conformance with applicable laws and regulations.

procurement lead time. The interval between the initiation of a procurement action and receipt of the products or services purchased as the result of such action.

profile dip. *Synonym* index dip.

profile dispersion. *See* dispersion.

profile parameter (*g*). In the power-law index profile of an optical fiber, the parameter, *g*, that defines the shape of the refractive-index profile. *Note:* The optimum value of g for minimum dispersion is approximately 2.

pro forma message. A standard form of message, that has elements that usually are understood by prearrangement among the originator, the addressee, and the communications system operators.

program. 1. A plan or routine for solving a problem on a computer. *Note:* Processing may include the use of an assembler, a compiler, an interpreter, or a translator to prepare the program for execution, as well as the execution of the program. The sequence of instructions may include statements and necessary declarations. 2. A sequence of instructions used by a computer to do a particular job or solve a given problem. 3. To design, write, and test programs.

program architecture. For a computer program, (a) the structure, relationships, and arrangement of the components of the program, (b) the program interfaces, and (c) the interface requirements for the program operating environment.

programmable. Pertaining to a device that can accept instructions that alter its basic functions.

programmable logic array (PLA). An array of gates having interconnections that can be programmed to perform a specific logical function.

programmable read-only memory (PROM). A storage device that, after being written to once, becomes a read-only memory.

programmer. 1. The part of digital equipment that controls the timing and sequencing of operations. 2. A person who prepares computer programs, *i.e.*, writes sequences of instructions for execution by a computer.

programming language. An artificial language that is used to generate or to express computer programs. *Note:* The language may be a high-level language, an assembly language, or a machine language. *See figure at* assembly language.

programming system. One or more programming languages and the software necessary for using these languages with particular automatic data processing equipment.

program origin. *See* computer program origin.

PROM. *Acronym for* programmable read-only memory.

prompt. 1. In interactive display systems, a message on the display surface of a display device to help the user to plan and execute subsequent operations. *Note:* Examples of prompts include (a) a blinking message displayed on a screen to inform the system operator of the status, condition, or mode the system is in and requiring the operator to take some action, and (b) a message that the system is ready to accept a command. 2. In a computer, communications, or data processing system, to inform a user that the system is ready for the next command, data element, or other input.

propagation. The motion of waves through or along a medium. *Note:* For electromagnetic waves, propagation may occur in a vacuum as well as in material media.

propagation constant. For an electromagnetic field mode varying sinusoidally with time at a given frequency, the logarithmic rate of change, with respect to distance in a given direction, of the complex amplitude of any field component. *Note:* The propagation constant, λ, is a complex quantity given by $\lambda = \alpha + i\beta$, where α, the real part, is the attenuation constant and β, the imaginary part, is the phase constant.

propagation mode. The manner in which radio signals travel from a transmitting antenna to a receiving antenna, such as ground wave, sky wave, direct wave, ground reflection, or scatter.

propagation path obstruction. A man-made or natural physical feature that lies near enough to a radio path to cause a measurable effect on path loss, exclusive of reflection effects. *Note:* An obstruction may lie to the side, above, or below the path. Ridges, bridges, cliffs, buildings, and trees are examples of obstructions. If the clearance from the nearest anticipated path position, over the expected range of Earth radius k-factor, exceeds 0.6 of the first Fresnel zone radius, the feature is not normally considered an obstruction.

propagation time delay. The time required for a signal to travel from one point to another.

proprietary standard. Documentation by a commercial entity specifying equipment, practices, or operations unique to that commercial entity.

proration. 1. The proportional distribution or allocation of parameters, such as noise power and transmission losses, among a number of tandem-connected items, such as equipment, cables, links, or trunks, in order to balance the performance of communications circuits. *Synonym* **budgeting.** 2. In a telephone switching center, the distribution or allocation of equipment or components proportionally among a number of functions, to provide a requisite grade of service.

protected distribution system (PDS). [A] wireline or fiber-optics telecommunication system that includes terminals and adequate acoustical, electrical, electromagnetic, and physical safeguards to permit its use for the unencrypted transmission of classified information. *Note:* A complete protected distribution system includes the subscriber and terminal equipment and the interconnecting lines. *Deprecated synonym* approved circuit.

protected frequency. A frequency that is not to be deliberately jammed by friendly forces, usually during a specified period.

protection. *Synonym* lockout (def. #5).

protection interval (PI). In high-frequency (HF) radio automatic link establishment, the period between changes in the time-of-day portion of the time-varying randomization data used for encrypting transmissions.

protection ratio. The minimum value of the wanted-to-unwanted signal ratio, usually expressed in decibels, at the receiver input determined under specified conditions such that a specified reception quality of the wanted signal is achieved at the receiver output.

protector. In telecommunications systems, a device used to protect facilities and equipment from abnormally high voltages or currents. *Note 1:* A protector may contain arresters. *Note 2:* Protectors may be designed to operate on short-duration phenomena, or long-duration phenomena. The duration should be specified.

protocol. 1. A formal set of conventions governing the format and control of interaction among communicating functional units. *Note:* Protocols may govern portions of a network, types of service, or administrative procedures. For example, a data link protocol is the specification of methods whereby data communications over a data link are performed in terms of the particular transmission mode, control procedures, and recovery procedures. 2. In layered communications system architecture, a formal set of procedures that are adopted to facilitate functional interoperation within the layered hierarchy.

protocol-control information. 1. The queries and replies among communications equipment to determine the respective capabilities of each end of the communications link. 2. For layered systems, information exchanged between entities of a given layer, via the service provided by the next lower layer, to coordinate their joint operation.

protocol converter. A functional unit that uses a specified algorithm to translate a bit stream from one protocol to another, for interoperation.

protocol data unit (PDU). 1. Information that is delivered as a unit among peer entities of a network and that may contain control information, address information, or data. 2. In layered systems, a unit of data that is specified in a protocol of a given layer and that consists of protocol-control information of the given layer and possibly user data of that layer.

protocol hierarchy. In open systems architecture, the distribution of network protocol among the various layers of the network.

protocol translator. In a communications system, the collection of hardware, software, firmware, or any combination of these, that is required or used to convert the protocols used in one network to those used in another network.

prototype. 1. A pre-production, functioning specimen(s) that is the first of its type, typically used for the evaluation of design, performance, and/or production potential. 2. A model suitable for evaluation of design, performance, and production potential.

provisioning. The act of acquiring telecommunications service from the submission of the requirement through the activation of service. *Note 1:* Provisioning includes all associated transmission, wiring, and equipment. *Note 2:* In NS/EP telecommunication services, *"provisioning"* and *"initiation"* are synonymous and include altering the state of an existing priority service or capability.

PS. *Abbreviation for* permanent signal.

pseudo bit-error ratio. *See* PBER.

pseudorandom noise. Noise that satisfies one or more of the standard tests for statistical randomness. *Note 1:* Although it seems to lack any definite pattern, pseudorandom noise contains a sequence of pulses that repeat themselves, albeit after a long time or a long sequence of pulses. *Note 2:* For example, in spread-spectrum systems, modulated carrier transmissions appear as pseudorandom noise to a receiver (a) that is not locked on the transmitter frequencies or (b) that is incapable of correlating a locally generated pseudorandom code with the received signal.

pseudorandom number generator. 1. A device that produces a stream of unpredictable, unbiased, and usually independent bits. 2. In cryptosystems, a random bit generator used for key generation or to start all the crypto-equipment at the same point in the key stream.

pseudorandom number sequence. 1. An ordered set of numbers that has been determined by some defined arithmetic process but is effectively a random number sequence for the purpose for which it is required. 2. A sequence of numbers that satisfies one or more of the standard tests for statistical randomness. *Note:* Although a pseudorandom number sequence appears to lack any definite pattern, it will repeat after a very long time interval or after a very long sequence of numbers.

PSK. *Abbreviation for* phase-shift keying.

PSN. *Abbreviation for* public switched network.

psophometer. An instrument that provides a visual indication of the audible effects of disturbing voltages of various frequencies. *Note:* A psophometer usually incorporates a weighting network. The characteristics of the weighting network depend on the type of circuit under investigation, such as whether the circuit is used for high-fidelity music or for normal speech.

psophometrically weighted dBm. *See* dBm(psoph), dBm0p.

******psophometric voltage.** Circuit noise voltage measured with a psophometer that includes a CCIF-1951 weighting network. *Note 1:* *"Psophometric voltage"* should not be confused with *"psophometric emf,"* *i.e.*, the emf in a generator or line with 600Ù internal resistance. For practical purposes, the psophometric emf is twice the corresponding psophometric voltage. *Note 2:* Psophometric voltage readings, *V*, in millivolts, are commonly converted to dBm(psoph) by $\text{dBm(psoph)} = 20 \log_{10} V - 57.78$.

psophometric weighting. A noise weighting established by the International Consultative Committee for Telephony (CCIF, which became CCITT and, more recently, ITU-T), designated as CCIF-1951 weighting, for use in a noise measuring set or psophometer. *Note:* The shape of this characteristic is virtually identical to that of F1A weighting. The psophometer is, however, calibrated with a tone of 800 Hz, 0 dBm, so that the corresponding voltage across 600 ohms produces a reading of 0.775 V. This introduces a 1-dBm adjustment in the formulas for conversion with dBa.

PSTN. *Abbreviation for* public switched telephone network.

PTF. *Abbreviation for* patch and test facility.

PTM. *Abbreviation for* pulse-time modulation.

PTT. *Abbreviation for* postal, telegraph, and telephone (organization). In countries having nationalized telephone and telegraph services, the organization, usually a governmental department, which acts as its nation's common carrier.

PTTI. *Abbreviation for* precise time and time interval.

public correspondence. Any telecommunication which the offices and stations must, by reason of their being at the disposal of the public, accept for transmission.

public data network (PDN). A network established and operated by a telecommunications administration, or a recognized private operating agency, for the specific purpose of providing data transmission services for the public.

public data transmission service. A data transmission service that is established and operated by a telecommunication administration, or a recognized private operating agency, and uses a public data network. *Note:* A public data transmission service may include circuit-switched, packet-switched, and leased-circuit data transmission.

public key cryptography. The type of cryptography in which the encryption process is publicly available and unprotected, but in which a part of the decryption key is protected so that only a party with knowledge of both parts of the decryption process can decrypt the cipher text. *Note:* Commonly called non-secret encryption in professional cryptologic circles. FIREFLY is an application of public key cryptography.

public land mobile network (PLMN). A network that is established and operated by an administration or by a recognized operating agency (ROA) for the specific purpose of providing land mobile telecommunications services to the public. *Note:* A PLMN may be considered as an extension of a fixed network, *e.g.* the Public Switched Telephone Network (PSTN) or as an integral part of the PSTN.

public switched network (PSN). Any common carrier network that provides circuit switching among public users. *Note:* The term is usually applied to public switched telephone networks, but it could be applied more generally to other switched networks, *e.g.*, packet-switched public data networks.

public switched NS/EP telecommunications services. Those NS/EP telecommunications services utilizing public switched networks. *Note:* Public switched NS/EP telecommunication services may include both interexchange and intraexchange network facilities (*e.g.*, switching systems, interoffice trunks, and subscriber loops).

public switched telephone network (PSTN). A domestic telecommunications network usually accessed by telephones, key telephone systems, private branch exchange trunks, and data arrangements. *Note:* Completion of the circuit between the call originator and call receiver in a PSTN requires network signaling in the form of dial pulses or multifrequency tones.

public utilities commission (PUC). In the United States, a state regulatory body charged with regulating intrastate utilities, including telecommunications systems. *Note:* In some states this regulatory function is performed by public service commissions or state corporation commissions.

PUC. *Abbreviation for* public utilities commission.

pull-in frequency range. The maximum frequency difference between the local oscillator or clock and the reference frequency of a phase-locked loop over which the local oscillator can be locked.

pulsating direct current. A direct current (dc) that changes in value at regular or irregular intervals. *Note:* A pulsating direct current may change in value, *i.e.,* be always present but at different levels, or it may be a current that is interrupted completely at regular or irregular intervals, but when present, is always in the same direction.

pulse. 1. A rapid, transient change in the amplitude of a signal from a baseline value to a higher or lower value, followed by a rapid return to the baseline value. 2. A rapid change in some characteristic of a signal, *e.g.,* phase or frequency, from a baseline value to a higher or lower value, followed by a rapid return to the baseline value.

pulse-address multiple access (PAMA). The ability of a communication satellite to receive signals from several Earth terminals simultaneously and to amplify, translate, and relay the signals back to Earth, based on the addressing of each station by an assignment of a unique combination of time and frequency slots. *Note:* This ability may be restricted by allowing only some of the terminals access to the satellite at any given time.

pulse amplitude. The magnitude of a pulse parameter, such as the field intensity, voltage level, current level, or power level. *Note 1:* Pulse amplitude is measured with respect to a specified reference and therefore should be modified by qualifiers, such as "average," "instantaneous," "peak," or "root-mean-square." *Note 2:* Pulse amplitude also applies to the amplitude of frequency- and phase-modulated waveform envelopes. *See illustration under* pulse.

pulse-amplitude modulation (PAM). Modulation in which the amplitude of individual, regularly spaced pulses in a pulse train is varied in accordance with some characteristic of the modulating signal. *Note:* The amplitude of the amplitude-modulated pulses conveys the information.

pulse broadening. An increase in pulse duration. *Note:* Pulse broadening may be specified by the impulse response, the root-mean-square pulse broadening, or the full-duration-at-half-maximum pulse broadening.

pulse carrier. An electromagnetic wave that (a) consists of a series of pulses usually of constant length, amplitude, spacing, and repetition rate when not modulated and (b) usually is used as a subcarrier.

pulse-code modulation (PCM). Modulation in which a signal is sampled, and the magnitude (with respect to a fixed reference) of each sample is quantized and digitized for transmission over a common transmission medium. *Note 1:* In conventional PCM, before being digitized, the analog data may be processed (*e.g.*, compressed), but once digitized, the PCM signal is not subjected to further processing (*e.g.*, digital compaction) before being multiplexed into the aggregate data stream. *Note 2:* PCM pulse trains may be interleaved with pulse trains from other channels.

pulse decay time. *Synonym* fall time.

pulse distortion. *See* distortion.

pulse duration. 1. In a pulse waveform, the interval between (a) the time, during the first transition, that the pulse amplitude reaches a specified fraction (level) of its final amplitude, and (b) the time the pulse amplitude drops, on the last transition, to the same level. *Note:* The interval between the 50% points of the final amplitude is usually used to determine or define pulse duration, and this is understood to be the case unless otherwise specified. Other fractions of the final amplitude, *e.g.*, 90% or 1/e (where e =

2.71828. . .), may also be used, as may the root-mean-square (rms) value of the pulse amplitude. *See illustration under* pulse. *Deprecated synonyms* **pulse** length, pulse width. 2. In radar, measurement of pulse transmission time in microseconds, that is, the time the radar's transmitter is energized during each cycle.

pulse-duration modulation (PDM). Modulation in which the duration of pulses is varied in accordance with some characteristic of the modulating signal. *Deprecated synonyms* pulse-length modulation, pulse-width modulation.

pulse duty factor. In a periodic pulse train, the ratio of the pulse duration to the pulse period.

pulse-frequency modulation (PFM). Modulation in which the pulse repetition rate is varied in accordance with some characteristic of the modulating signal. *Note:* Pulse-frequency modulation is analogous to frequency modulation of a carrier wave, in which the instantaneous frequency is a continuous function of the modulating signal.

pulse-interval modulation. *See* pulse-position modulation.

pulse length. *Deprecated synonym for* pulse duration.

pulse-length modulation *Deprecated synonym for* pulse-duration modulation.

pulse link repeater (PLR). A device that interfaces concatenated E&M signal paths. *Note 1:* A PLR converts a ground, received from the E lead of one signal path, to –48 Vdc, which is applied to the M lead of the concatenated signal path. *Note 2:* In many commercial carrier systems, the channel bank cards or modules have a "PLR" option that permits the direct connection, *i.e.,* concatenation, of E&M signaling paths, without the need for separate PLR equipment.

pulse period. The reciprocal of the pulse repetition rate.

pulse-position modulation (PPM). Modulation in which the temporal positions of the pulses are varied in accordance with some characteristic of the modulating signal.

pulse-repetition frequency (PRF). In radar applications, *synonym* pulse repetition rate.

pulse repetition rate. The number of pulses per unit time.

pulse-repetition-rate modulation. *Synonym* pulse-frequency modulation.

pulse rise time. *See* rise time.

pulse string. *Synonym* pulse train.

pulse stuffing. *See* bit stuffing.

pulse-time modulation (PTM). The general class of pulse-code modulation in which the time of occurrence of some characteristic of the pulsed carrier is varied with respect to some characteristic of the modulating signal. *Note:* PTM includes pulse-position modulation and pulse-duration modulation.

pulse train. A series of pulses having similar characteristics. *Synonym* pulse string.

pulse width. *Deprecated synonym for* pulse duration.

pulse-width modulation (PWM). *Deprecated synonym for* pulse-duration modulation.

pulsing. In telephony, the transmission of address information to a switching office by means of pulses, *i.e.,* signals, that originate from the subscriber, *i.e.,* user, equipment. *Note:* Examples of pulsing methods are dual-tone multifrequency (DTMF) signaling, in which a unique pair of audio frequencies represents each of the respective numerals or other characters on a keypad, and rotary dialing, in which dc pulses are generated by a rotary dial. *Synonyms* key pulsing (when using a keypad), dial pulsing (when using a rotary dial).

pump frequency. The frequency of an oscillator used to provide sustaining power to a device, such as a laser or parametric amplifier, that requires rf or optical power.

pumping. The action of an oscillator that provides cyclic inputs to an oscillating reaction device. *Note:* Examples of pumping are the action that results in amplification of a signal by a parametric amplifier, and the action that provides a laser or maser with an input signal at the appropriate frequency to sustain stimulated emission.

pure binary numeration system. *See* binary notation.

pushbutton dialing. Dialing in which (a) pushbuttons or keys are used to actuate and connect audible tone oscillators to a line, (b) each button or key corresponds to a unique frequency or set of frequencies, and (c) each pushbutton or key represents a unique digit or symbol.

push-down file. *See* last-in first-out.

push-to-talk (PTT) operation. In telephone or two-way radio systems, that method of communication over a speech circuit in which the talker is required to keep a switch operated while talking. *Note:* In two-way radio, push-to-talk operation must be used when the same frequency is employed by both transmitters. For use in noisy environments, or for privacy, some telephone handsets have push-to-talk switches that allow the speaker to be heard only when the switch is activated. *Synonym* press-to-talk operation.

push-to-type operation. In telegraph or data transmission systems, that method of communication in which the operator at a station must keep a switch operated in order to send messages. *Note 1:* Push-to-type operation is used in radio systems where the same frequency is employed for transmission and reception. *Note 2:* Push-to-type operation is a derivative form of transmission and may be used in simplex, half-duplex, or duplex operation. *Synonym* press-to-type operation.

PVC. *Abbreviation for* permanent virtual circuit.

pW. *Abbreviation for* picowatt. A unit of power equal to 10^{-12} W, *i.e.*, –90 dBm. *Note:* One picowatt is usually used as a reference level for both weighted and unweighted noise measurements. The type of measurement must be specified.

PWM. *Abbreviation for* pulse-width modulation, *which is a deprecated synonym for* pulse-duration modulation.

pWp. *Abbreviation for* picowatt, psophometrically weighted. *See* noise weighting.

pWp0. *Abbreviation for* picowatts, psophometrically weighted, measured at a zero-dBm transmission level point. *See* dBm(psoph), psophometer.

pX. *Abbreviation for* peak envelope power (of a radio transmitter).

PX. 1. *Abbreviation for* private exchange; *See* PBX. 2. *Abbreviation for* peak envelope power.

p x 64. In video teleconferencing, pertaining to a family of CCITT Recommendations, where p is a non-zero positive integer indicating the number of 64 kb/s channels. *Note:* The p x 64 family includes CCITT Recommendations H.261, H.221, H.242, H.230, and H.320. These Recommendations form the basis for video telecommunications interoperability.

QA. *Abbreviation for* quality assurance.

QAM. *Abbreviation for* quadrature amplitude modulation.

QC. *Abbreviation for* quality control.

QOS. *Abbreviation for* quality of service.

QPSK. *Abbreviation for* quadrature phase-shift keying.

quad. A group of four wires composed of two pairs twisted together. *Note:* The pairs have a fairly long length of twist and the quad a fairly short length of twist.

quadded cable. A cable formed of multiples of quads, paired and separately insulated, and contained under a common jacket.

quadratic profile. *Synonym* parabolic profile.

quadrature. 1. The state of being separated in phase by 90° (ð/2 radians). 2. Pertaining to the phase relationship between two periodic quantities varying with the same period, that is, with the same frequency or repetition rate, when the phase difference between them is one-quarter of their period.

quadrature amplitude modulation (QAM). Quadrature modulation in which the two carriers are amplitude modulated.

quadrature modulation. Modulation using two carriers out of phase by 90° and modulated by separate signals.

quadrature phase-shift keying (QPSK). Phase-shift keying in which four different phase angles are used. *Note:* In QPSK, the four angles are usually out of phase by 90°. *Synonyms* quadriphase, quaternary phase-shift keying.

quadriphase. *Synonym* quadrature phase-shift keying.

quadruple diversity. In radio communication, diversity transmission and reception in which four independently fading signals are used. *Note:* Quadruple diversity may be accomplished through the use of space, frequency, angle, time, or polarization multiplexing, or combinations of these.

quadruply clad fiber. A single-mode optical fiber that has four claddings. *Note 1:* Each cladding has a refractive index lower than that of the core. With respect to one another, their relative refractive indices are, in order of distance from the core, lowest, highest, lower, higher. *Note 2:* A quadruply clad fiber has the advantage of very low macrobending losses. It also has two zero-dispersion points, and moderately low dispersion over a wider wavelength range than a singly clad fiber or a doubly clad fiber. *See illustration under* refractive index profile.

quality assurance (QA). 1. All actions taken to ensure that standards and procedures are adhered to and that delivered products or services meet performance requirements. 2. The planned systematic activities necessary to ensure that a component, module, or system conforms to established technical requirements. 3. The policy, procedures, and systematic actions established in an enterprise for the purpose of providing and maintaining a specified degree of confidence in data integrity and accuracy throughout the lifecycle of the data, which includes input, update, manipulation, and output.

quality control (QC). A management function whereby control of the quality of (a) raw materials, assemblies, produced materiel, and components, (b) services related to production, and (c) management, production, and inspection processes is exercised for the purpose of preventing undetected production of defective materiel or the rendering of faulty services.

quality factor. In a reactive circuit, the ratio of the reactance in ohms divided by the resistance in ohms.

quality of service (QOS). 1. The performance specification of a communications channel or system. *Note:* QOS may be quantitatively indicated by channel or system performance parameters, such as signal-to-noise ratio (S/N), bit error ratio (BER), message throughput rate, and call blocking probability. 2. A subjective rating of telephone communications quality in which listeners judge transmissions by qualifiers, such as excellent, good, fair, poor, or unsatisfactory.

quantization. A process in which the continuous range of values of an analog signal is sampled and divided into nonoverlapping (but not necessarily equal) subranges, and a discrete, unique value is assigned to each subrange. *Note:* An application of quantization is its use in pulse-code modulation. If the sampled signal value falls within a given subrange, the sample is assigned the corresponding discrete value for purposes of modulation and transmission.

quantization error. *Synonym* quantizing distortion.

quantization level. In the quantization process, the discrete value assigned to a particular subrange of the analog signal being quantized.

quantization noise. *Synonym* quantizing noise.

quantized feedback. In a digital feedback loop, the digital signal that is fed back. *Note 1:* Several forms of analog-to-digital converters contain a quantized feedback loop following the basic A-D converter. *Note 2:* The feedback signal is often processed before introducing it to the loop.

quantizing distortion. Distortion that results from the quantization process. *Synonym* **quantization error.**

quantizing levels. In digital transmission, the number of discrete signal levels transmitted as the result of signal digitization.

quantizing noise. Noise caused by the error of approximation in quantization. *Note:* Quantizing noise is dependent on the particular quantization process used and the statistical characteristics of the quantized signal. *Synonym* quantization noise.

quantum efficiency. In an optical source or detector, the ratio of the number of output quanta to the number of input quanta. *Note:* Input and output quanta need not both be photons.

quantum-limited operation. *Synonym* quantum-noise-limited operation.

quantum noise. Noise attributable to the discrete and probabilistic nature of physical phenomena and their interactions. *Note 1:* Quantum noise represents the fundamental limit of the achievable signal-to-noise ratio of an optical communication system. This limit is never achieved in practice. *Note 2:* Examples of quantum noise are photon noise in an optical signal and shot noise in an electrical conductor or semiconductor.

quantum-noise-limited operation. Operation wherein the minimum detectable signal is limited by quantum noise. *Synonym* quantum-limited operation.

quarter common intermediate format (QCIF). A video format defined in CCITT Recommendation H.261 that is characterized by 176 luminance pixels on each of 144 lines, with half as many chrominance pixels in each direction. *Note:* QCIF has one-fourth as many pixels as the full common intermediate format.

quartz clock. A clock containing a quartz oscillator that determines the accuracy and precision of the clock.

quartz oscillator. An oscillator in which a quartz crystal is used to stabilize the frequency. *Note:* The piezoelectric property of the quartz crystal results in a nearly constant output frequency, which is dependent upon the crystal size, shape, and excitation.

quasi-analog signal. A digital signal that has been converted to a form suitable for transmission over a specified analog channel. *Note:* The specification of the analog channel should include frequency range, bandwidth, signal-to-noise ratio, and envelope delay distortion. When quasi-analog form of signaling is used to convey message traffic over dial-up telephone systems, it is often referred to as voice-data. A modem may be used for the conversion process.

quasi-analog transmission. Transmission in which a special-purpose modulator is used to convert digital signals into an analog form suitable for transmission over an analog voice-grade circuit. *Note:* A complementary demodulator is used to recover the digital signal at the other end of the circuit. *See* modem.

quaternary phase-shift keying. *Synonym* quadrature phase-shift keying.

quaternary signal. A digital signal having four significant conditions.

query call. In adaptive high-frequency (HF) radio, an automatic-link-establishment call that requests responses from stations having connectivity to the destination specified in the call.

queue. A set of items, such as telephone calls or packets, arranged in sequence. *Note:* Queues are used to store events occurring at random times and to service them according to a prescribed discipline that may be fixed or adaptive.

queue traffic. 1. A series of outgoing or incoming calls waiting for service. 2. In a store-and-forward switching center, the outgoing messages awaiting transmission at the outgoing line position.

queuing. The process of entering elements into or removing elements from a queue.

queuing delay. 1. In a switched network, the time between the completion of signaling by the call originator and the arrival of a ringing signal at the call receiver. *Note:* Queues may be caused by delays at the originating switch, intermediate switches, or the call receiver servicing switch. 2. In a data network, the sum of the delays between the request for service and the establishment of a circuit to the called data terminal equipment (DTE). 3. In a packet-switched network, the sum of the delays encountered by a packet between the time of insertion into the network and the time of delivery to the addressee.

queuing theory. The theoretical study of waiting lines, expressed in mathematical terms—including components such as number of waiting lines, number of servers, average wait time, number of queues or lines, and probabilities of queue times' either increasing or decreasing. *Note:* Queuing theory is directly applicable to network telecommunications, server queuing, mainframe computer queuing of telecommunications terminals, and advanced telecommunications systems.

quieting. In an FM receiver, the phenomenon that results in less noise when an unmodulated carrier is present than when there is no carrier present. *Note:* Quieting is expressed in dB.

raceway. Within a building, an enclosure, *i.e.*, channel, used to contain and protect wires, cables, or bus bars.

rack. A frame upon which one or more units of equipment are mounted. *Note:* DOD racks are always vertical.

racon. *See* radar beacon.

rad. *Acronym for* radiation absorbed dose. The basic unit of measure for expressing absorbed radiant energy per unit mass of material. *Note 1:* A rad corresponds to an absorption of 0.01 J/kg, *i.e.*, 100 ergs/g. *Note 2:* The absorbed radiant energy heats, ionizes, and/or destroys the material upon which it is incident.

rad.. *Abbreviation for* radian(s).

radar. *Acronym for* radio detection and ranging. 1. A radio detection system that transmits short bursts (pulses) of rf energy and detects their echos from objects (targets) such as aircraft or ships. *Note:* The round-trip propagation time for the echo return may be used to determine the target's range (distance from the radar's antenna). If the transmitting antenna has a narrow beam (the usual case), the azimuth or elevation of the target may also be determined. *Synonym* primary radar. *Contrast with* secondary radar. 2. A radio detection device that provides information on range, azimuth, and/or elevation of objects. 3. A radiodetermination system based on the comparison of reference signals with radio signals reflected, or retransmitted, from the position to be determined.

radar beacon (racon). 1. A transmitter-receiver associated with a fixed navigational mark which, when triggered by a radar, automatically returns a distinctive signal which can appear on the display of the triggering radar, providing range, bearing and identification information. 2. A receiver-transmitter combination which sends out a coded signal when triggered by the proper type of pulse, enabling determination of range and bearing information by the interrogating station or aircraft.

radar blind range. The range that corresponds to the situation in which a radar transmitter is on and hence the receiver must be off, so that the radar transmitted signal does not saturate, *i.e.*, does not blind, its own receiver. *Note:* Radar blind ranges occur because there is a time interval between transmitted pulses that corresponds to the time required for a pulse to propagate to the object, *i.e.*, to the target, and its reflection to travel back. This causes an attempt to measure the range just as the radar transmitter is transmitting the next pulse. However, the receiver is off, therefore this particular range cannot be measured. The width of the range value that cannot be measured depends on the duration of the time that the radar receiver is off, which depends on the duration of the transmitted pulse. The return-time interval could be coincident with the very next radar-transmitted pulse, *i.e.*, the first pulse following a transmitted pulse, or the second, or the third, and so on, giving rise to a succession of blind ranges. The blind ranges are given by $r_m = (mc)/(2fn)$, where r_m is the blind range for a given value of m, m is a positive integer that indicates which of the blind ranges is being determined, c is the velocity of electromagnetic wave propagation in vacuum (approximately 3×10^8 m/s), f is the radar pulse repetition rate, and n is the refractive index of the transmission medium (nearly 1 for air). The radar blind range is independent of the radar radio frequency (rf) of the radar pulse.

radar blind speed. The magnitude of the radial component of velocity of an object, *i.e.*, a target, relative to a radar site, that cannot be measured by the radar unit. *Note:* Radar blind speeds occur because of the relationship between the transmitted pulse repetition rate (PRR) and the received pulse-repetition rate. The Doppler pulse repetition rate is the difference between the transmitted and received pulse repetition rates. For example, when the object is stationary with respect to the radar site, the reflected PRR is the same as the transmitted PRR and therefore a net zero signal is indicated for the radial component of velocity. If it happens that the Doppler PRR is the same as the transmitted PRR, *i.e.*, the illuminating PRR, or it is a multiple of the transmitted PRR, a zero signal is also obtained and hence the radar is blind to these speeds, one for each multiple of the transmitted pulse repetition rate. It is not the absolute magnitude of the speed of the object that is measured, but only the radial component of the speed. The radial components of blind speeds, v_m, are given by $v_m = m\lambda f/102$, where v is the blind speed in knots, m is the multiple of the radar pulse repetition rate and the number of the blind speed, namely a positive integer, 1, 2, 3, 4, . . ., for the first, second, third, fourth, and so on, blind speed, λ is the wavelength of the illuminating radar in centimeters; f is the transmitter pulse repetition rate in pps (pulses per second); and the 102 is a units conversion factor.

radar cross section. An expression of the extent to which an object, *i.e.,* a target, reflects radar pulses, usually with respect to their point of origin. *Note:* The radar cross section of an aircraft can vary by a factor of over 100, depending on the aspect angle of the aircraft to the radar transmitter. Radar reflection off the nose of the aircraft usually represents the smallest radar cross section, while a broadside presentation to the signal produces the greatest cross section. Shape, surface roughness, and reflective material as well as orientation also affect the radar cross section.

radar intelligence (RADINT). Intelligence derived from data collected by radar.

radar line-of-sight (LOS) equation. An equation that expresses the radar horizon range (RHR), given by

$$RHR_s = \sqrt{2h} + \sqrt{2a}$$

$$= 1.414\left(\sqrt{h} + \sqrt{a}\right),$$

where RHR_s is the radar horizon range in statute miles, h is the antenna height in feet, and a is the object critical altitude, *i.e.,* the target altitude in feet, below which the radar cannot illuminate the object. *Note:* The RHR is also given by

$$RHR_k = 1.42\left(\sqrt{h} + \sqrt{a}\right),$$

where RHR_k is the radar horizon range in kilometers when h and a are in meters. The effective Earth radius, namely 4/3 times the actual Earth radius, is used in deriving these formulas. The effective Earth radius for LOS varies with carrier frequency. Second-order differentials are neglected. They contribute less than 0.1%.

radar mile. The time required for a radar pulse to travel 1 mile (~1.6 km) to an object, *i.e.*, to a target; reflect; and return to the receiver. *Note:* A radar statute mile is approximately 10.8 μs (microseconds); a radar nautical mile is approximately 12.4 μs. The time for any other radar unit distance is readily determined, such as the radar meter or the radar kilometer.

radar resolution cell. The volume of space that is occupied by a radar pulse and that is determined by the pulse duration and the horizontal and vertical beamwidths of the transmitting radar. *Note:* The radar cannot distinguish between two separate objects that lie within the same resolution cell. The radar resolution cell depth (*RCD*) remains constant regardless of the distance from the transmitting antenna. It does not increase with range. The *RCD* is given by $RCD = 150d$, where the *RCD* is in meters and *d* is the pulse duration in microseconds. The height of the cell and the width of the cell do increase with range. These are given by $W = (HBW)(R/57)$ and $H = (VBW)(R/57)$, where *W* is the width of the cell, *HBW* is the horizontal beamwidth in degrees, *R* is the range, *H* is the height of the cell, and *VBW* is the vertical beamwidth in degrees. The range, *R*, is the distance from the radar antenna to the reflecting object, *i.e.*, the target. The width and height will come out in the same units in which the range is given. For example, if the range is given in meters, the width and height of the radar resolution cell will be in meters. The 57 merely converts degrees to radians. If the beamwidths are given in radian measure, the 57 is omitted.

radar signature. 1. The detailed waveform of a detected radar echo. *Note:* Radar signatures may be used to identify or distinguish among objects, *i.e.*, targets, such as aircraft, decoys, missiles with warheads, and chaff. 2. The detailed characteristics of a radar transmission.

Note: Radar signatures based upon emission analysis may be used to identify or distinguish among specific radar types.

RADHAZ. *Acronym for* electromagnetic radiation hazards.

radian (rad.). A unit of plane angle measure equal to the angle subtended at the center of a circle by an arc equal in length to the radius of the circle. *Note:* One radian is equal to 360°/2π, which is approximately 57° 17' 44.6".

radiance. Radiant power, in a given direction, per unit solid angle per unit of projected area of the source, as viewed from the given direction. *Note:* Radiance is usually expressed in watts per steradian per square meter.

radiant emittance. Radiant power emitted into a full sphere, *i.e.,* 4ð sr (steradians), by a unit area of a source, expressed in watts per square meter. *Synonym* radiant exitance.

radiant energy. Energy in the form of electromagnetic waves. *Note 1:* Radiant energy may be calculated by integrating radiant power with respect to time. *Note 2:* Radiant energy is usually expressed in joules.

radiant exitance. *Synonym* radiant emittance.

radiant flux. *Deprecated synonym for* radiant power.

radiant intensity. Radiant power per unit solid angle, usually expressed in watts per steradian.

radiant power. The rate of flow of electromagnetic energy, *i.e.,* radiant energy. *Note 1:* Radiant power is usually expressed in watts, *i.e.,* joules per second. *Note 2:* The modifier is often dropped and *"power"* is used to mean *"radiant power"*. *Deprecated synonyms* flux, radiant flux.

radiation. 1. In communication, the emission of energy in the form of electromagnetic waves. 2. The outward flow of energy from any source in the form of radio waves.

radiation angle. In fiber optics, half the vertex angle of the cone of light emitted at the exit face of an optical fiber. *Note:* The cone boundary is usually defined (a) by the angle at which the far-field irradiance has decreased to a specified fraction of its maximum value or (b) as the cone within which there is a specified fraction of the total radiated power at any point in the far field. *Synonym* output angle.

radiation efficiency. At a given frequency, the ratio of the power radiated to the total power supplied to the radiator.

radiation field. *Synonym* far-field region.

radiation-hardened fiber. An optical fiber made with core and cladding materials that recover, within a specified period of time, a specified percentage of their intrinsic transparency after darkening from exposure to a radiation pulse.

radiation mode. For an optical fiber, an unbound mode. *Note:* In an optical fiber, a radiation mode is one having fields that are transversely oscillatory everywhere external to the waveguide, and which exists even at the limit of zero wavelength. Specifically, a radiation mode is one for which

$$\beta = \sqrt{n^2(a)k^2 - (l/a)^2} \quad ,$$

where β is the imaginary part (phase term) of the axial propagation constant, integer ℓ is the azimuthal index of the mode, $n(a)$ is the refractive index, where a is the core radius, and k is the free-space wave number, $k = 2\pi/\lambda$, where λ is the wavelength. Radiation modes correspond to refracted rays in the terminology of geometric optics. *Synonym* unbound mode.

radiation pattern. 1. The variation of the field intensity of an antenna as an angular function with respect to the axis. *Note:* A radiation pattern is usually represented graphically for the far-field conditions in either horizontal or vertical plane. 2. In fiber optics, the relative power distribution at the output of a fiber or active device as a function of position or angle. *Note 1:* The near-field radiation pattern describes the radiant emittance ($W \bullet m^{-2}$) as a function of position in the plane of the exit face of an optical fiber. *Note 2:* The far-field radiation pattern describes the irradiance as a function of angle in the far-field region of the exit face of an optical fiber. *Note 3:* The radiation pattern may be a function of the length of the fiber, the manner in which it is excited, and the wavelength. *Synonym* directivity pattern.

radiation resistance. The resistance that, if inserted in place of an antenna, would consume the same amount of power that is radiated by the antenna.

radiation scattering. The diversion of radiation (thermal, electromagnetic, or nuclear) from its original path as a result of interaction or collisions with atoms, molecules, or larger particles in the atmosphere or other media between the source of radiation (*e.g.*, a nuclear explosion) and a point some distance away. As a result of scattering, radiation (especially gamma rays and neutrons) will be received at such a point from many directions instead of only from the direction of the source.

RADINT. *Acronym for* radar intelligence.

radio. 1. Telecommunication by modulation and radiation of electromagnetic waves. 2. A transmitter, receiver, or transceiver used for communication via electromagnetic waves. 3. A general term applied to the use of radio waves.

radio altimeter. Radionavigation equipment, on board an aircraft or spacecraft, used to determine the height of the aircraft or the spacecraft above the Earth's surface or another surface.

radio and wire integration (RWI). The combining of wire circuits with radio facilities.

radio baseband. *See* baseband.

radiobeacon station. A station in the radionavigation service the emissions of which are intended to enable a mobile station to determine its bearing or direction in relation to the radiobeacon station.

radio beam. A radiation pattern from a directional antenna, such that the energy of the transmitted electromagnetic wave is confined to a small angle in at least one dimension.

radio channel. An assigned band of frequencies sufficient for radio communication. *Note 1:* The bandwidth of a radio channel depends upon the type of transmission and the frequency tolerance. *Note 2:* A channel is usually assigned for a specified radio service to be provided by a specified transmitter.

radio common carrier (RCC). A common carrier engaged in the provision of Public Mobile Service, which is not also in the business of providing landline local exchange telephone service. These carriers were formerly called *"miscellaneous common carriers."*

radiocommunication. Telecommunication by means of radio waves.

radiocommunication service. A service as defined in this Section [of the *Radio Regulations*] involving the transmission, emission and/or reception of radio waves for specific telecommunication purposes. In these regulations, unless otherwise stated, any radiocommunication service relates to terrestrial radiocommunication.

radio control. The remote control of an apparatus by signals conveyed by electromagnetic waves. *Note:* Radio control may be used to control the movement of an aircraft, vehicle, missile, or other mobile unit, either manned or unmanned, from a radio station on the ground or in another mobile unit.

radio-coverage diagram. A diagram that shows the area within which a radio station is broadcasting an effective signal strength in relation to a given standard. *Note:* An example of a radio-coverage diagram is a polar plot, in each direction from the antenna, of the distance from the antenna at which the signal strength is equal to a specified value, *i.e.,* it is the locus of all points at which the signal strength is equal to a specified value.

radio detection. The detection of the presence of an object by radiolocation without precise determination of its position.

radio detection and ranging. *See* radar.

radiodetermination. The determination of the position, velocity and/or other characteristics of an object, or the obtaining of information relating to these parameters, by means of the propagation properties of radio waves.

radiodetermination-satellite service. A radiocommunication service for the purpose of radiodetermination involving the use of one or more space stations. This service may also include feeder links necessary for its own operation.

radiodetermination station. A station in the radiodetermination service.

radio direction-finding [RDF]. Radiodetermination using the reception of radio waves for the purpose of determining the direction of a station or object.

radio direction-finding station. A radiodetermination station using radio direction-finding.

radio equipment. As defined in *Federal Information Management Regulations*, any equipment or interconnected system or subsystem of equipment (both transmission and reception) that is used to communicate over a distance by modulating and radiating electromagnetic waves in space without artificial guide. This does not include such items as microwave, satellite, or cellular telephone equipment.

radio fadeout. *See* flutter.

radio field intensity. *Synonym* field strength.

radio fix. 1. The locating of a radio transmitter by bearings taken from two or more direction finding stations, the site of the transmitter being at the point of intersection. 2. The location of a ship or aircraft by determining the direction of radio signals coming to the ship or aircraft from two or more sending stations, the locations of which are known.

radio frequency (rf). Any frequency within the electromagnetic spectrum normally associated with radio wave propagation. *For designation of subdivisions, see electromagnetic spectrum and associated diagram.*

radio frequency assignment. *See* frequency assignment.

radio frequency channel assignment. *Synonym* frequency assignment.

radio frequency interference (RFI). *Synonym* electromagnetic interference.

radio horizon. The locus of points at which direct rays from an antenna are tangential to the surface of the Earth. *Note:* If the Earth were a perfect sphere and there were no atmospheric anomalies, the radio horizon would be a circle. In practice, the distance to the radio horizon is affected by the height of the transmitting antenna, the height of the receiving antenna, atmospheric conditions, and the presence of obstructions, *e.g.,* mountains.

radio horizon range (RHR). The distance at which a direct radio wave can reach a receiving antenna of given height from a transmitting antenna of given height. *Note:* The radio horizon range in nautical miles, R, is given by the relation $R = 1.23(h_t^{1/2} + h_r^{1/2})$, where h_t and h_r are the heights of the transmitting and receiving antennas in feet. The radio horizon range, R, in nautical miles is also given by the relation $R = 2.23(h_t^{1/2} + h_r^{1/2})$, where h_t and h_r are the heights of the transmitting and receiving antennas in meters. The effective Earth radius, 4/3 times the actual Earth radius, is used in deriving the formulae. Second-order differentials are neglected. They are of the order of 0.1%.

radio interface. The common boundary between a mobile station and the radio equipment in the network, which is the boundary defined by functional characteristics, physical interconnection characteristics, signal characteristics, and other characteristics as appropriate.

radiolocation. Radiodetermination used for purposes other than those of radionavigation.

radiolocation land station. A station in the radiolocation service not intended to be used while in motion.

radiolocation mobile station. A station in the radiolocation service intended to be used while in motion or during halts at unspecified points.

radiolocation service. A radiodetermination service for the purpose of radiolocation.

radiological monitoring. *Synonym* monitoring (def. #4).

radiometry. The science of radiation measurement. *See Table of Radiometric Terms.*

radionavigation. 1. Radio-location intended for the determination of position or direction or for obstruction warning in navigation. 2. Radiodetermination used for the purposes of navigation, including obstruction warning.

radionavigation land station. A station in the radionavigation service not intended to be used while in motion.

radionavigation mobile station. A station in the radionavigation service intended to be used while in motion or during halts at unspecified points.

radionavigation-satellite service. A radiodetermination-satellite service used for the purpose of radionavigation. This service may also include feeder links necessary for its operation.

radio net. 1. An organization of radio stations that is capable of direct communication on a common frequency. 2. An organization of radio stations that broadcast common programming, not necessarily simultaneously, at different frequencies from different locations.

radio paging. The use of a pocket-size radio receiver capable of alerting its wearer that there is a phone call, either from a displayed phone number or to a predesignated number. *Note:* Radio paging may be considered a subset of paging. *Synonym* beeping.

radio personal terminal. *See* personal terminal.

radio range. 1. The distance from a transmitter at which the signal strength remains above the minimum usable level for a particular antenna and receiver combination. 2. A radio aid to air navigation that creates an infinite number of paths in space throughout a given sector or azimuth angle by various methods of transmission and reception of electromagnetic waves.

radio range station. A radionavigation land station in the aeronautical radionavigation service providing radial equisignal zones. (In certain instances a radio range station may be placed on board a ship.)

radio recognition and identification. *See* identification, friend or foe.

Radio Regulations Board. A permanent organization of the International Telecommunication Union (ITU) that implements frequency assignment policy and maintains the Master International Frequency Register (MIFR). *Note: Formerly* International Frequency Registration Board (IFRB).

radio relay. 1. The reception and retransmission by a radio station of signals that are received either from another radio station or from a wire, fiber optic, microwave, coaxial cable, or other link of an integrated land line and radio communications system component. 2. A terrestrial point-to-point communications system, such as a microwave-relay communications system or a satellite communications system. *Note:* The siting of radio-relay stations and the radio coverage diagrams of the antenna patterns are arranged for minimum interference with satellite

Earth stations. The analog and digital baseband arrangements are similar to satellite systems. Radio-relay links may form part of the connection between an Earth station and a switching center.

radio relay system. A point-to-point radio transmission system in which signals are received, conditioned, and retransmitted by one or more intermediate radio stations.

radiosonde. An automatic radio transmitter in the meteorological aids service usually carried on an aircraft, free balloon, kite, or parachute, and which transmits meteorological data.

radiotelegram. A telegram, originating in or intended for a mobile station or a mobile Earth station transmitted on all or part of its route over the radiocommunication channels of the mobile service or of the mobile-satellite service.

radio telegraphy. The transmission of telegraphic codes by means of radio.

radiotelemetry. Telemetry by means of radio waves.

radiotelephone call. A telephone call, originating in or intended for a mobile station or a mobile Earth station, transmitted on all or part of its route over the radiocommunication channels of the mobile service or of the mobile-satellite service.

radiotelephone distress frequency. An international distress and calling frequency for mobile radiotelephone stations, survival craft, and emergency position-indicating radio beacons. *Note:* An example of a radiotelephone distress frequency is 2180 kHz.

radio telephony. The transmission of speech by means of modulated radio waves.

radio teletypewriter (RTTY). A teletypewriter employed in a communication system using radio circuits. *Note:* Such systems are spoken of as RATT systems.

radiotelex call. A telex call, originating in or intended for a mobile station or a mobile Earth station, transmitted on all or part of its route over the radiocommunication channels of the mobile service or the mobile-satellite service.

radio wave. An electromagnetic wave of a frequency arbitrarily lower than 3000 GHz. *Synonym* Hertzian wave.

radio watch shift. *Synonym* area broadcast shift.

radio-wire integration. *See* radio and wire integration.

RAM. *Acronym for* random access memory.

Raman amplifier. *Synonym* fiber amplifier.

Raman scattering. The generation of many different wavelengths of light from a nominally single-wavelength source (a) by means of lasing action and interaction with molecules, thereby creating many different excited molecular energy levels that will produce photons of various energy levels, *i.e.,* various wavelengths, when transitions to lower excited states occur and (b) by the beating together of two frequencies, thus inducing dipole moments in molecules at the difference frequencies and thereby causing modulation of laser-molecule interaction, which, in turn, produces light at side frequencies, *i.e.,* side wavelengths relative to the nominal wavelength.

random access discrete address (RADA). A communications technique in which radio users share one wide frequency band instead of each user's being assigned a narrow band.

random access memory (RAM). A read/write, nonsequential-access memory used for the storage of instructions and data. *Note 1:* RAM access time is essentially the same for all storage locations. *Note 2:* RAM is characterized by a shorter access time than disk or tape storage. *Note 3:* RAM is usually volatile.

randomizer. 1. A device used to invert the sense of pseudorandomly selected bits of a bit stream to avoid long sequences of bits of the same sense. *Note:* The same selection pattern must be used on the receive terminal in order to restore the original bit stream. 2. [An] analog or digital source of unpredictable, unbiased, and usually independent bits. *Note:* Randomizers can be used for several different functions, including key generation or to provide a starting state for a key generator.

random noise. Noise consisting of a large number of transient disturbances with a statistically random time distribution. *Note:* Thermal noise is an example of random noise.

random number. 1. A number selected from a known set of numbers in such a way that each number in the set has the same probability of occurrence. 2. A number obtained by chance. 3. One of a sequence of numbers considered appropriate for satisfying certain statistical tests or believed to be free from conditions that might bias the result of a calculation.

range. *See* radio range.

ranging. The measurement of the distance to a remote object (target), from a known observation or reference point. *Note:* Ranging may be accomplished by geometric means, *e.g.,* triangulation, or by the measurement of the transit time of an electromagnetic or acoustic signal. Ranging has application to navigation and cartography.

raster. A predetermined pattern of scanning lines within a display space. *Note:* An example of a raster is the pattern followed by an electron beam scanning the screen of a television camera or receiver.

raster count. The total number of raster scanning lines within a display space.

raster density. In display systems, the number of scanning lines per unit distance perpendicular to the scanning direction.

raster scanning. Scanning in which the motion of the scanning spot follows a raster.

rated output power. That power available at a specified output of a device under specified conditions of operation. *Note:* Rated output power may be further described; *e.g.,* maximum rated output power, average rated output power.

ratio-squared combiner. *Synonym* maximal-ratio combiner.

ray. A geometric representation of a lightwave by a line normal to the electromagnetic wavefront; *i.e.,* in the direction of propagation of the wave.

Rayleigh distribution. A mathematical statement, usually applied to frequency distributions of random variables, for the case in which two orthogonal variables are independent and normally distributed with unit variance.

Rayleigh fading. In electromagnetic wave propagation, phase-interference fading caused by multipath, and which may be approximated by the Rayleigh distribution.

Rayleigh scattering. Of an electromagnetic wave propagating in a material medium, scattering caused by refractive-index inhomogeneities that are small compared to the wavelength. *Note 1:* Rayleigh scattering losses vary as the reciprocal of the fourth power of the wavelength. *Note 2:* Ionospheric scattering is caused partly by Rayleigh scattering.

ray optics. *Synonym* geometric optics.

RBOC. *Acronym for* Regional Bell Operating Company.

RC. *Abbreviation for* reflection coefficient.

RCC. *Abbreviation for* radio common carrier.

RDF. *Abbreviation for* radio direction finding.

read head. A magnetic head capable of reading only.

reading. The acquisition or interpretation of data from a storage device, from a data medium, or from another source.

read-only memory (ROM). A memory in which data, under normal conditions, can only be read. *Synonym* nonerasable storage.

read-only storage. A storage device in which the contents cannot be modified, except by a particular user, or when operating under particular conditions, *e.g.*, a storage device in which writing is prevented by a lockout. *Synonym* fixed storage.

read/write opening. *Synonym* read/write slot.

read/write slot. An opening in the jacket of a diskette to allow access to the read/write heads. *Synonym* read/write opening.

ready-for-data signal. 1. A call-control signal that is transmitted by the data circuit-terminating equipment (DCE) to the data terminal equipment (DTE) to indicate that the connection is available for data transfer between both DTEs. 2. A signal that (a) is sent in the backward direction in the interexchange data channel, to indicate that all the succeeding exchanges involved in the connection have been through-connected, or (b) is sent in the forward direction in the interexchange data channel to indicate that all the preceding exchanges involved in the connection have been through-connected. *Note:* The ready-for-data signal is sent by the user terminal. It corresponds to the ready-for-data state at the user interface.

real power. *See* effective power.

real time. 1. The actual time during which a physical process occurs. 2. Pertaining to the performance of a computation during the actual time that the related physical process occurs, in order that results of the computation can be used in guiding the physical process.

reasonableness check. A test to determine whether a value conforms to specified criteria. *Note:* A reasonableness check can be used to eliminate questionable data points from subsequent processing. *Synonym* wild-point detection.

receive-after-transmit time delay. The time interval between (a) the instant of keying off the local transmitter to stop transmitting and (b) the instant the local receiver output has increased to 90% of its steady-state value in response to an rf signal from a distant transmitter. *Note 1:* The rf signal from the distant transmitter must exist at the local receiver input prior to, or at the time of, keying off the local transmitter. *Note 2:* Receive-after-transmit time delay applies only to half-duplex operation.

received noise power. 1. The calculated or measured noise power, within the bandwidth being used, at the receive end of a circuit, channel, link, or system. 2. The absolute power of the noise measured or calculated at a receive point. *Note:* The related bandwidth and the noise weighting must also be specified. 3. The value of noise power, from all sources, measured at the line terminals of telephone set's receiver. *Note:* Either flat weighting or some other specific amplitude-frequency characteristic or noise weighting characteristic must be associated with the measurement.

received signal level (RSL). The signal level at a receiver input terminal. *Note 1:* The signal bandwidth and the established reference level must be specified. *Note 2:* The RSL is usually expressed in dB with respect to 1 mW, *i.e.,* 0 dBm.

receive only (RO). Pertaining to a device or a mode of operation capable of receiving messages, but not of transmitting messages.

receiver attack-time delay. The time interval from (a) the instant a step rf signal, at a level equal to the receiver threshold of sensitivity, is applied to the receiver input to (b) the instant the receiver output amplitude reaches 90% of its steady-state value. *Note:* If a squelch circuit is operating, the receiver attack-time delay includes the time for the receiver to break squelch.

receiver lockout system. *Synonym* lockout (def. #4).

receiver release-time delay. The time interval from removal of rf energy at the receiver input until the receiver output is squelched.

recognized operating agency (ROA). Any operating agency, as defined in the ITU Convention (Geneva, 1992), which operates a public correspondence or broadcasting service and upon which the obligations provided for in Article 6 of the ITU Constitution are imposed by the Member in whose territory the head office of the agency is situated, or by the Member which has authorized this operating agency to establish and operate a telecommunication service on its territory. *Formerly* recognized private operating agency (RPOA).

reconditioned carrier reception. *Synonym* exalted-carrier reception.

reconstructed sample. An analog sample generated at the output of a decoder when a specified character signal is applied at its input. *Note:* The amplitude of the reconstructed sample is proportional to the value of the corresponding encoded sample.

record. 1. A set of data treated as a unit. 2. To write data on a medium, such as magnetic tape, magnetic disk, or optical disk.

record communication. 1. A telecommunications process that produces an electronic message that is transmitted, received, stored or archived, and may be retrieved. 2. A telecommunications process, that produces a hard copy record of the transmission, such as a teletypewriter printout or a facsimile printout.

recorder warning tone. A half-second burst of 1400 Hz applied to a telephone line every 15 seconds to indicate to the called party that the calling party is recording the conversation. *Note:* The recorder warning tone is required by law to be generated as an integral part of any recording device used for the purpose and is required to be not under the control of the calling party. The tone is recorded together with the conversation.

record information. All forms (*e.g.*, narrative, graphic, data, computer memory) of information registered in either temporary or permanent form so that it can be retrieved, reproduced, or preserved.

recording density. *Synonym* bit density.

recording spot. In a facsimile recorder, the spot that is used to generate the recorded copy on the record medium.

record medium. 1. The physical medium on which information is stored in recoverable form. 2. In facsimile transmission, the physical medium on which the recorder forms an image of the object, *i.e.,* creates the recorded copy. *Note:* The record medium and the record sheet may be identical. *Synonym* record sheet.

record sheet. *Synonym* record medium.

record traffic. 1. Traffic that is recorded, in permanent or quasipermanent form, by the originator, the addressee, or both. 2. Traffic that is permanently or semipermanently recorded in response to administrative procedures or public law.

recovery. In a database management system, the procedures and capabilities available for reconstruction of the contents of a database to a state that prevailed before the detection of processing errors and before the occurrence of a hardware or software failure that resulted in the destruction of some or all of the stored data.

recovery procedure. 1. The actions necessary to restore an automated information system's data files and computational capability after a system failure. 2. In data communications, a process whereby a data station attempts to resolve conflicting or erroneous conditions arising during the transfer of data.

RED/BLACK concept. [The] separation of electrical and electronic circuits, components, equipment, and systems that handle classified plain text (RED) information, in electrical signal form, from those which handle unclassified (BLACK) information in the same form.

RED signal. 1. [A] Telecommunications or automated information system signal that would divulge classified information if recovered and analyzed. 2. In cryptographic systems, a signal containing classified information that has NOT been encrypted.

reduced carrier single-sideband emission. A single-sideband emission in which the degree of carrier suppression enables the carrier to be reconstituted and to be used for demodulation.

reduced carrier transmission. A form of amplitude-modulation in which the carrier is transmitted at a controlled level below that which is required for demodulation, but at a level sufficient to serve as a frequency reference.

redundancy. 1. In the transmission of data, the excess of transmitted message symbols over that required to convey the essential information in a noise-free circuit. *Note:* Redundancy may be introduced intentionally (as in the case of error detection or correction codes) or inadvertently (such as by oversampling a band-limited signal, inefficient formats, *etc.*). 2. In a communication system, surplus capability usually provided to improve the reliability and quality of service.

redundancy check. 1. A method of verifying that any redundant hardware or software in a communication system is in an operational condition. 2. A check that uses one or more extra binary digits or characters attached to data for the detection of errors.

redundant code. A code using more signal elements than necessary to represent the intrinsic information. *Note:* The redundancy may be used for error-control purposes.

reference antenna. An antenna that may be real, virtual, or theoretical, and has a radiation pattern that can be used as a basis of comparison with other antenna radiation patterns. *Note:* Examples of reference antennas are unit dipoles, half-wave dipoles, and isotropic, *i.e.,* omnidirectional antennas.

reference black level. [In television,] The level corresponding to the specified maximum excursion of the luminance signal in the black direction.

reference circuit. A hypothetical circuit of specified equivalent length and configuration, and having a defined transmission characteristic or characteristics, used primarily as a reference for measuring the performance of other, *i.e.,* real, circuits or as a guide for planning and engineering of circuits and networks. *Note:* Normally, several types of reference circuits are defined, with different configurations, because communications are required over a wide range of distances. A group of related reference circuits is also called a *reference system.*

reference clock. 1. A clock with which another clock is compared. 2. A clock, usually of high stability and accuracy, used to govern a network of mutually synchronized clocks of lower stability. *Note:* The failure of a reference clock does not necessarily cause loss of synchronism.

reference configuration. In ISDN, a combination and arrangement of functional groups and reference points that reflect possible network topology.

reference frequency. 1. A standard fixed frequency from which operational frequencies may be derived or with which they may be compared. *Note:* The reference frequency may be used to specify an assigned frequency or fix a characteristic or carrier frequency. 2. A frequency having a fixed and specific position with respect to the assigned frequency. The displacement of this frequency with respect to the assigned frequency has the same absolute value and sign that the displacement of the characteristic frequency has with respect to the center of the frequency band occupied by the emission.

reference monitor. [An] access control concept that refers to an abstract machine that mediates all accesses to objects by subjects.

reference noise. The magnitude of circuit noise chosen as a reference for measurement. *Note:* Many different levels with a number of different weightings are in current use, and care must be taken to ensure that the proper parameters are stated. *See* dBa, dBa(F1A), dBa(HA1), dBa0, dBm, dBm(psoph), dBm0, dBrn, dBrnC, dBrnC0, dBrn(f_1-f_2), dBrn(144-line), dBx.

reference point. In ISDN, a logical point between two, nonoverlapping functional groups. *Note:* When equipment is placed at a reference point, that reference point is designated an interface.

reference surface. In optical-fiber technology, that surface of an optical fiber that is used to contact the transverse-alignment elements of a component such as a connector or mechanical splice. *Note:* For telecommunications-grade fibers, the reference surface is the outer surface of the cladding. For plastic-clad silica (PCS) fibers, which have a strippable polymer cladding (not to be confused with the polymer overcoat of an all-glass fiber), the reference surface may be the core.

reference system. A group of related reference circuits.

reference transmission level point. *See* relative transmission level, transmission level point.

reference white level. [In television,] The level corresponding to the specified maximum excursion of the luminance signal in the white direction.

reflectance. The ratio of reflected power to incident power, generally expressed in dB or percent.

reflected code. *Synonym* Gray code.

reflecting layer. In the ionosphere, a layer that has a free-electron density sufficient to reflect radio waves. *Note:* The principal reflecting layers are the E, F_1, and F_2 layers in the daylight hemisphere. *Note 2:* A critical frequency is associated with the reflection by each layer.

reflecting loss. *See* **reflection loss.**

reflection. The abrupt change in direction of a wave front at an interface between two dissimilar media so that the wave front returns into the medium from which it originated. *Note 1:* Reflection may be specular (*i.e.,* mirror-like) or diffuse (*i.e.,* not retaining the image, only the energy) according to the nature of the interface. *Note 2:* Depending on the nature of the interface, *i.e.,* dielectric-conductor or dielectric-dielectric, the phase of the reflected wave may or may not be inverted.

reflection coefficient (RC). 1. The ratio of the amplitude of the reflected wave and the amplitude of the incident wave. 2. At a discontinuity in a transmission line, the complex ratio of the electric field strength of the reflected wave to that of the incident wave. *Note 1:* The reflection coefficient may also be established using other field or circuit quantities. *Note 2:* The reflection coefficient is given by

$$RC = \left| \frac{Z_1 - Z_2}{Z_1 - Z_2} \right| = \frac{SWR - 1}{SWR + 1},$$

where Z_1 is the impedance toward the source, Z_2 is the impedance toward the load, the vertical bars designate absolute magnitude, and *SWR* is the standing wave ratio.

reflection loss. 1. At a discontinuity or impedance mismatch, *e.g.*, in a transmission line, the ratio of the incident power to the reflected power. *Note 1:* Reflection loss is usually expressed in dB. *Note 2:* The reflection loss, L_r, is given by

$$L_r = 20 \log_{10} \left| \frac{Z_1 - Z_2}{Z_1 + Z_2} \right| = 10 \log_{10} \frac{(Z_1 - Z_2)^2}{(Z_1 + Z_2)^2},$$

where Z_1 and Z_2 are the respective impedances, and the vertical bars designate absolute magnitude. 2. In an optical fiber, the loss that takes place at any discontinuity of refractive index, especially at an air-glass interface such as a fiber endface, at which a fraction of the optical signal is reflected back toward the source. *Note:* This reflection phenomenon is also called *"Fresnel reflection loss,"* or simply, *"Fresnel loss."* At normal incidence, the fraction of reflected power is expressed by the formula

$$L_f = 10 \log_{10} \frac{(n_1 - n_2)^2}{(n_1 + n_2)^2},$$

where n_1 and n_2 are the respective indices of refraction.

reflective array antenna. An antenna, such as a billboard antenna, in which the driven elements are situated at a predetermined distance from a surface designed to reflect the signal in a desired direction. *Note:* Reflective array antennas (a) usually have many driven elements working in conjunction with an electrically large reflecting surface to produce a unidirectional beam, (b) may be used to increase antenna gain, *i.e.*, reduce radiation in unwanted directions, and (c) may contain parasitic elements as well as driven elements.

reflectivity. The reflectance at the surface of a material so thick that the reflectance does not change with increasing thickness, *i.e.,* the intrinsic reflectance of the surface, irrespective of other parameters such as the reflectance of the rear surface. *Note:* The term *"reflectivity"* is no longer in common use. *See* reflectance.

reflector. In an antenna, one or more conducting elements or surfaces that reflect incident radiant energy.

refracted ray. 1. A ray that undergoes a change of velocity, or in the general case, both velocity and direction, as a result of interaction with the material medium in which it travels. 2. In an optical fiber, a ray that is refracted from the core into the cladding. Specifically a ray having direction such that

$$\frac{n^2(r) - n^2(a)}{1 - (r/a)^2 \cos^2\phi(r)} \leq \sin^2\theta(r),$$

where r is the radial distance from the fiber axis, $\phi(r)$ is the azimuthal angle of projection of the ray at r on the transverse plane, $\phi(r)$ is the angle the ray makes with the fiber axis, $n(r)$ is the refractive index at r, $n(a)$ is the refractive index at the core radius, a. Refracted rays correspond to radiation modes in the terminology of mode descriptors.

refraction. Retardation, and–in the general case–redirection, of a wavefront passing through (a) a boundary between two dissimilar media or (b) a medium having a refractive index that is a continuous function of position, *e.g.,* a graded-index optical fiber. *Note:* For two media of different refractive indices, the angle of refraction is closely approximated by Snell's Law.

refraction profile. *Synonym* refractive index profile.

refractive index (n, η). Of a medium, the ratio of the velocity of propagation of an electromagnetic wave in vacuum to its velocity in the medium. *Synonym* index of refraction.

refractive index contrast. In an optical fiber, a measure of the relative difference in refractive index of the core and cladding. *Note:* Refractive index contrast, Δ, is given by $\Delta = (n_1^2 - n_2^2)/(2n_1^2)$, where n_1 is the maximum refractive index in the core and n_2 is the refractive index of the homogeneous cladding.

refractive index profile. Of the cross section of an optical fiber, the description, *i.e.*, plot, of the value of the refractive index as a function of distance from the fiber axis along a diameter. *Synonyms* index profile, refraction profile.

reframing time. The time interval between the instant at which a valid frame-alignment signal is available at the receiving data terminal equipment and the instant at which frame alignment is established. *Note:* The reframing time includes the time required for replicated verification of the validity of the frame-alignment signal. *Synonym* frame-alignment recovery time.

refresh. To reproduce, repeatedly, a display image on a display surface, so that the image remains visible.

regeneration. 1. In a regenerative repeater, the process by which digital signals are amplified, reshaped, retimed, and retransmitted. *Synonym* positive feedback. 2. In a storage or display device, the restoration of stored or displayed data that have deteriorated. *Note:* For example, conventional cathode-ray tube displays must be continually regenerated for the data to remain displayed. 3. In computer graphics, the sequence of events needed to generate a display image from its representation in storage.

regenerative feedback. Feedback in which the portion of the output signal that is returned to the input has a component that is in phase with the input signal.

regenerative repeater. A repeater, designed for digital transmission, in which digital signals are amplified, reshaped, retimed, and retransmitted. *Synonym* regenerator.

regenerator. *Synonym* regenerative repeater.

Regional Bell Operating Company (RBOC). One of the seven holding companies formed by divestiture by the American Telephone and Telegraph Company of its local Bell System operating companies, and to which one or more of the Bell System local telephone companies were assigned.

regional center. *See* office classification.

register. 1. A device, accessible to one or more input circuits, that accepts and stores data. *Note:* A register is usually used only as a device for temporary storage of data. 2. A temporary-memory device used to receive, hold, and transfer data (usually a computer word) to be operated upon by a processing unit. *Note:* Computers typically contain a variety of registers. General purpose registers may perform many functions, such as holding constants or accumulating arithmetic results. Special purpose registers perform special functions, such as holding the instruction being executed, the address of a storage location, or data being retrieved from or sent to storage.

registered jack (RJ). Any of the series of jacks, described in the *Code of Federal Regulations,* Title 47, part 68, used to provide interface to the public telephone network.

registration. 1. The accurate positioning of an entity relative to a reference. 2. *See* FCC registration program.

registration program. *See* FCC Registration Program.

rekeying. The changing of one or more keys that are used for either COMSEC or TRANSEC functions.

relative error. The ratio of an absolute error to the true, specified, or theoretically correct value of the quantity that is in error.

relative spectral width. *See* spectral width.

relative transmission level. The ratio of the signal power, at a given point in a transmission system, to a reference signal power. *Note:* The ratio is usually determined by applying a standard test tone at zero transmission level point (or applying adjusted test tone power at any other point) and measuring the gain or loss to the location of interest. A distinction should be made between the standard test tone power and the expected median power of the actual signal required as the basis for the design of transmission systems.

relay. 1. To retransmit a received message from one station to another station. 2. An electromechanical or semiconductor switch (*i.e.,* solid-state relay) in which a current or voltage applied across one port or terminal controls electrical currents or voltages that appear across another terminal or terminals.

relay configuration. An operating configuration in which a circuit is established between two stations via an intermediate relay station. Two links are used simultaneously and the channel connections at the relay station are accomplished completely within the station.

relay station. An intermediate station that passes information between terminals or other relay stations.

released loop. *Synonym* switched loop.

release time. 1. The time interval between (a) the instant that an enabling signal (as in a vogad or echo suppressor) is discontinued, and (b) the instant at which suppression ceases. 2. The time interval between (a) the instant a relay coil is de-energized, and (b) the instant that contact closure ceases (or, depending on the nature of the relay, is established).

reliability. 1. The ability of an item to perform a required function under stated conditions for a specified period of time. 2. The probability that a functional unit will perform its required function for a specified interval under stated conditions. 3. The continuous availability of communication services to the general public, and emergency response activities in particular, during normal operating conditions and under emergency circumstances with minimal disruption.

reliability assessment. 1. The process of determining whether existing hardware, firmware, or software has achieved a specified level of operational reliability. 2. The process of determining the achieved level of operational reliability of existing hardware, firmware, or software. *Synonym* **reliability evaluation.**

reliability evaluation. *Synonym* reliability assessment.

remote access. 1. Pertaining to communication with a data processing facility from a remote location or facility through a data link. 2. A PABX service feature that allows a user at a remote location to access by telephone PABX features, such as access to wide area telephone service (WATS) lines. *Note:* For remote access, individual authorization codes are usually required.

remote access data processing. Data processing in which some input/output functions are performed by devices that are connected to a computer system by means of data communication.

remote batch entry. Submission of batches of data through an input unit that has access to a computer through a data link.

remote batch processing. Batch processing in which input-output units have access to a computer through a data link.

remote call forwarding. A service feature that allows calls coming to a remote call-forwarding number to be automatically forwarded to any answering location designated by the call receiver. *Note:* Customers may have a remote-forwarding telephone number in a central switching office without having any other local telephone service in that office.

remote clock. 1. A clock that is remote from a particular facility, such as a communications station or node, with which it is associated. 2. A clock that is remote from another clock to which it is to be compared.

remote control equipment. Devices used to perform monitoring, controlling, and/or supervisory functions, at a distance.

remote job entry (RJE). In computer operations, a mode of operation that allows execution of job instructions received from a remote site and return of the output to the same or a different remote site via a communications link.

remote login. A login that allows a user terminal to connect to a host computer via a network or direct telecommunications link, and to interact with that host computer as if the user terminal were directly connected to that host computer. *Synonym* remote logon.

remote logon. *Synonym* remote login.

remote operations service element (ROSE) protocol. A protocol that (a) provides remote operation capabilities, (b) allows interaction between entities in a distributed application, and (c) upon receiving a remote operations service request, allows the receiving entity to attempt the operation and report the results of the attempt to the requesting entity.

remote orderwire. An extension of a local orderwire to a point convenient for personnel to perform required operational and maintenance functions.

remote rekeying. The encrypted transmission of keys from a remote source.

remote trunk arrangement (RTA). Arrangement that permits the extension of TSPS functions to remote locations.

REN. *Acronym for* ringer equivalency number.

reorder tone. *See* busy signal.

repeater. 1. An analog device that amplifies an input signal regardless of its nature, *i.e.,* analog or digital. 2. A digital device that amplifies, reshapes, retimes, or performs a combination of any of these functions on a digital input signal for retransmission. *Note:* The term *"repeater"* originated with telegraphy and referred to an electromechanical device used to regenerate telegraph signals. Use of the term has continued in telephony and data communications.

repeating coil. A voice-frequency transformer characterized by a closed core, a pair of identical balanced primary (line) windings, a pair of identical but not necessarily balanced secondary (drop) windings, and low transmission loss at voice frequencies. *Note:* It permits transfer of voice currents from one winding to another by magnetic induction, matches line and drop impedances, and prevents direct conduction between the line and the drop.

repeat-request (RQ) system. *Synonym* ARQ.

reperforator. In teletypewriter systems, a device used to punch a tape in accordance with arriving signals, permitting reproduction of the signals for retransmission.

repertory dialer. A telephone set that stores a group of numbers frequently called by a customer and transmits the dialing information to the central office by a single action. *Contrast with* **speed calling.**

reproducibility. *Synonym* precision (def. #1, #2, #3).

reproducible fault. In computer and telecommunications systems, a fault that will occur each time the set of conditions causing the fault occurs. *Note:* The conditions under which the fault occurs and the precision with which the fault occurs must be specified when determining the reproducibility of a fault.

reproduction speed. 1. In facsimile systems, the rate at which recorded copy is produced. *Note:* The reproduction speed is usually expressed (a) as the area of recorded copy produced per unit time, such as square meters per second or (b) as the number of pages per minute. 2. In duplicating equipment, the rate at which copies are made. *Note:* The reproduction speed is usually expressed in pages per minute.

request data transfer. A signal sent by the DTE to the DCE to request the establishment of a data connection.

request-repeat (RQ) signal. A signal from a receiver to a transmitter asking that a message be transmitted again.

request-to-send signal. A signal that is generated by a receiver in order to condition a remote transmitter to commence transmission.

reradiation. 1. Radiation, at the same or different wavelengths, *i.e.,* frequencies, of energy received from an incident wave. 2. Undesirable radiation of signals locally generated in a radio receiver. *Note:* Radiation might cause interference or reveal the location of the device.

rerouting. Recommencement of route selection from the first point of routing control, when congestion is encountered at some intermediate switching point in the connection that is to be established.

resale carrier. A company that redistributes the services of a commercial carrier and retails the services to the public.

resale service. In FCC deliberations and rulings, the right of a buyer of basic telecommunication services, such as private lines, foreign exchanges, or WATS, to resell and/or share with others the unused capacity.

reserve capacity. Installed capacity of a system which is not normally utilized but can be made available when required.

reserved circuit service. In ISDN applications, a telecommunications service that establishes a communication path at a preset time (requested by the user) in response to a user-network signaling request.

reserved word. In programming languages, a keyword whose definition is fixed by the programming language and which cannot be changed by the user. *Note:* In Ada® and COBOL all keywords are reserved words, while Fortran has no reserved words.

reset mode. The parameters initially programmed for basic operation.

resettability. A measure of the ability to duplicate controllable conditions. *Note:* An example of resettability is the ability to reset the frequency controls of radio equipment.

resident. Pertaining to computer programs that remain on a particular storage device.

residual error rate. *Synonym* undetected error ratio.

residual error ratio. 1. The ratio of (a) the number of bits, unit elements, characters, or blocks incorrectly received but undetected or uncorrected by the error-control equipment to (b) the total number of bits, unit elements, characters or blocks. 2. The error ratio that remains after attempts at correction are made.

residual modulation. *Synonym* carrier noise level.

resistance hybrid. A network of resistors to which four branches of a circuit may be connected to make them conjugate in pairs. *Note:* The primary use of a resistance hybrid is to convert between 2-wire and 4-wire communications circuits. Such conversion is necessary when repeaters are introduced in a 2-wire circuit.

resolution. 1. The minimum difference between two discrete values that can be distinguished by a measuring device. *Note:* High resolution does not necessarily imply high accuracy. 2. The degree of precision to which a quantity can be measured or determined. 3. A measurement of the smallest detail that can be distinguished by a sensor system under specific conditions.

resolving power. A measure of the ability of a lens or optical system to form separate and distinct images of two objects with small angular separation. *Note:* The resolving power of an optical system is ultimately limited by diffraction by the aperture. Thus, an optical system cannot form a perfect image of a point.

resonance. In an electrical circuit, the condition that exists when the inductive reactance and the capacitive reactance are of equal magnitude, causing electrical energy to oscillate between the magnetic field of the inductor and the electric field of the capacitor. *Note 1:* Resonance occurs because the collapsing magnetic field of the inductor generates an electric current in its windings that charges the capacitor and the discharging capacitor provides an electric current that builds the magnetic field in the inductor, and the process is repeated. *Note 2:* At resonance, the series impedance of the two elements is at a minimum and the parallel impedance is a maximum. Resonance is used for tuning and filtering, because resonance occurs at a particular frequency for given values of inductance and capacitance. Resonance can be detrimental to the operation of communications circuits by causing unwanted sustained and transient oscillations that may cause noise, signal distortion, and damage to circuit elements. *Note 3:* At resonance the inductive reactance and the capacitive reactance are of equal magnitude. Therefore, $|\omega L| = |1/\omega C|$, where $\omega = 2\pi f$, in which f is the resonant frequency in hertz, L is the inductance in henrys, and C is the capacity in farads when standard SI units are used. Thus,

$$f = \frac{\pi}{2\sqrt{LC}}.$$

resonant antenna. *Synonym* periodic antenna.

resonant cavity. *See* optical cavity.

resource controller (RC). The processor(s) that control access to satellite payload communications resources within an individual satellite program.

respond opportunity. In data transmission, the link level logical control condition during which a given secondary station may transmit a response.

response. 1. A reply to a query. 2. In data transmission, the content of the control field of a response frame advising the primary station concerning the processing by the secondary station of one or more command frames. 3. The effect of an active or passive device upon an input signal.

response frame. In data transmission, all frames that may be transmitted by a secondary station.

response PDU. A protocol data unit (PDU) transmitted by a logical link control (LLC) sublayer in which the PDU command/response (C/R) bit is equal to "1".

response time. The time a system or functional unit takes to react to a given input. *Note:* For example, in data processing, the response time perceived by the end user is the interval between (a) the instant at which an operator at a terminal enters a request for a response from a computer and (b) the instant at which the first character of the response is received at a terminal. In a data system, the system response time is the interval between the receipt of the end of transmission of an inquiry message and the beginning of the transmission of a response message to the station originating the inquiry.

response timer (T_K). In multilevel precedence and preemption, the device that controls the length of time that the call receiver of the precedence call has to accept the incoming precedence call. *Note:* The length of the time is usually set in the range of 4 s to 30 s.

responsiveness. Ability of an entity to provide service within the required time. *Note:* The term *timeliness* is sometimes used incorrectly to mean responsiveness.

responsivity. In a photodetector, the ratio of the electrical output to the optical input. *Note 1:* Responsivity is usually expressed in amperes per watt, or volts per watt, of incident radiant power. *Note 2:* Responsivity is a function of the wavelength of the incident radiation and the bandgap of the material of which the photodetector is made. *Deprecated synonym* sensitivity.

restart. The resumption of the execution of a computer program using the data recorded at a checkpoint.

restitution. A series of significant conditions determined by the decisions taken according to the products of the demodulation process.

restoration. Action taken to repair and return to service one or more telecommunications services, including repair of a damaged or impaired telecommunications facility, that have a degraded quality of service or have a service outage. *Note:* Restoration may be done by various means, such as patching, routing, substitution of component parts, or selecting other pathways.

restricted access. A class of service in which users may be denied access to one or more of the system features or operating levels.

restricted area. *Synonym* controlled space.

restricted channel. In digital communications systems, a channel that has a useful capacity of only 56 kb/s (kilobits per second), instead of 64 kb/s. *Note:* The restricted channel, currently common in North America, was originally developed to satisfy a ones-density limitation in T1 circuits.

retrieval service. In interactive telecommunications, a service allowing access to and retrieval of stored information, *e.g.,* the information within a database.

retrograde orbit. Of a satellite orbiting the Earth, an orbit in which the projection of the satellite's position on the (Earth's) equatorial plane revolves in the direction opposite that of the rotation of the Earth.

return loss. The ratio, at the junction of a transmission line and a terminating impedance or other discontinuity, of the amplitude of the reflected wave to the amplitude of the incident wave. *Note 1:* Return loss is usually expressed in dB. *Note 2:* Return loss is a measure of the dissimilarity between impedances in metallic transmission lines and loads, or between refractive indices in dielectric media, *e.g.*, optical fibers. *Note 3:* In a metallic transmission line, return loss is given by

$$L_r = 10 \log_{10} \left| \frac{Z_1 + Z_2}{Z_1 - Z_2} \right| ,$$

where Z_1 is the impedance toward the source and Z_2 is the impedance toward the load, and the vertical bars indicate magnitude. *Note 4:* For dielectric media, *e.g*, optical fibers, *see* reflection loss (def #2).

return-to-zero (RZ). A digital code having two information states, *e.g.*, "0" and "1" or "mark" and "space", in which code the signal returns to a rest state during a portion of the bit period.

reverse-battery signaling. Loop signaling in which battery and ground are reversed on the tip and ring of the loop to give an "off-hook" signal when the call receiver answers. *Note:* Reverse-battery signaling may be used either for a short period, or for the duration of a call, to indicate that it is a toll call.

revertive pulsing. In telephone networks, a means of controlling distant switching selections by pulsing. *Note:* In revertive pulsing, the near end receives signals from the far end.

rf. *Abbreviation for* radio frequency.

rf bandwidth. *See* occupied bandwidth, necessary bandwidth.

RFI. *Abbreviation for* radio frequency interference. *See* electromagnetic interference.

rf power margin. 1. The amount of transmitter power above that which is computed by the link designer as the minimum required to meet specified link performance. *Note:* The rf power margin allows for uncertainties in (a) empirical components of the signal level prediction method, (b) terrain characteristics, (c) atmospheric conditions, and (d) equipment performance parameters. 2. At any given time in an operational link, the reserve transmitter power over that which is required to maintain specified link performance.

rf tight. Offering a high degree of electromagnetic shielding effectiveness.

RGB. *Abbreviation for* red-green-blue. Pertaining to the use of three separate signals to carry the red, green, and blue components, respectively, of a color video image. *Note:* The image is not NTSC-encoded; RGB typically results in higher resolution than that specified by the National Television Standards Committee.

rhombic antenna. A directional antenna that is composed of long-wire radiators that form the sides of a rhombus, the two halves of which are fed equally in opposite phase at one apex. *Note:* A rhombic antenna is usually terminated at the apex opposite the driven apex, which termination makes its radiation pattern unidirectional. It is bidirectional if the opposite apex is unterminated.

RI. *Abbreviation for* routing indicator.

ribbon cable. 1. Any cable constructed as a ribbon with parallel elements. 2. A fiber optic cable in which the optical fibers are held in grooves and laminated within a flat semirigid strip of material, such as plastic, that positions, holds, and protects them. *Note:* Ribbon cables may be stacked to produce fiber optic cables with large numbers of fibers. Buffers, strength members, fillers, and jacketing are usually added to produce the final cable. [After 2196].

right-hand (or clockwise) polarized wave. An elliptically or circularly polarized wave, in which the electric field vector, observed in any fixed plane, normal to the direction of propagation, whilst looking in the direction of propagation, rotates with time in a right-hand or clockwise direction. *Synonym* clockwise polarized wave.

right-hand rule. *Synonym* Fleming's rule.

ring. 1. In telephony, a signal of specific duration and character that indicates to a user (customer, subscriber) that a calling party is engaged in an access attempt. 2. *Synonym* ring network. *See* network topology.

ringaround. 1. The improper routing of a call back through a switching center already engaged in attempting to complete the same call. 2. In secondary surveillance radar, the presence of false targets declared as a result of transponder interrogation by side lobes of the interrogating antenna.

ringback signal. 1. In telephony, a signal, usually consisting of an audio tone interrupted at a slow rate, provided to a caller to indicate that the called-party instrument is receiving a ringing signal. *Note:* This signal may be generated by the called-party servicing switch or by the calling-party switch. 2. A ringing signal returned to a caller to indicate that one of the types of delayed automatic calling is now ringing the called party.

ringback tone. *Synonym* audible ringing tone.

ringdown. In telephony, a method of signaling an operator in which telephone ringing current is sent over the line to operate a lamp and cause the drop of a self-locking relay. *Note 1:* Ringdown (a) is used in manual operation, as distinguished from dialing, (b) uses a continuous or pulsed ac signal transmitted over the line, and (c) may be used with or without a switchboard. *Note 2:* The term *"ringdown"* originated in magneto telephone signaling in which cranking the magneto in a telephone set would not only *"ring"* its bell but also cause a lever to fall *"down"* at the central office switchboard.

ringdown circuit. In telephony, a circuit in which manually generated signaling power is used to perform ringdown.

ringdown signaling. In telephony, the application of a signal to a line (a) to operate a line signal lamp or a supervisory signal lamp at a switchboard or (b) to ring a called receiver instrument.

ringer equivalency number (REN). A number determined in accordance with the *Code of Federal Regulations,* Title 47, part 68, which number represents the ringer loading effect on a line. *Note:* A ringer equivalency number of 1 represents the loading effect of a single traditional telephone set ringing circuit. Modern telephone instruments may have a REN lower than 1. The total REN expresses the total loading effect of the subscriber's equipment on the central office ringing current generator. The service provider usually sets a limit, *e.g.,* 3, 4, or 5 (representing "extension," *i.e.,* parallel-connected telephones), to the total REN on a subscriber's loop. The actual number of instruments across the loop may be greater than the service provider's REN limit, if their respective individual RENs are less than 1.

ring latency. In a ring network, such as a token ring network, the time required for a signal to propagate once around the ring. *Note 1:* Ring latency may be measured in seconds or in bits at the data transmission rate. *Note 2:* Ring latency includes signal propagation delays in (a) the ring medium, (b) the drop cables, and (c) the data stations connected to the ring network.

ring network. *See* network topology.

ring topology. *See* network topology.

ring transit time. *See* round-trip delay time.

R interface. For a basic rate access in an ISDN environment, the interfacing specifications covering pre-ISDN standards (*e.g.*, EIA-232C).

rip cord. Of an optical cable, a parallel cord of strong yarn that is situated under the jacket(s) of the cable for the purpose of facilitating jacket removal preparatory to splicing or breaking out. *Note:* The rip cord is exposed by carefully removing or severing a portion of the jacket near the end of the cable. It is then grasped with the fingers, or usually, with a tool such as a pair of pliers, and pulled to sever the jacket for the remainder of the desired distance.

ripple voltage. 1. In a dc voltage, the alternating component that is residually retained from rectification of ac power, or from generation and commutation. 2. In a dc voltage, the alternating component that is coupled into a circuit from a source of interference.

rise time. In the approximation of a step function, the time required for a signal to change from a specified low value to a specified high value. Typically, these values are 10% and 90% of the step height.

RJ. *Abbreviation for* registered jack.

RJE. *Abbreviation for* remote job entry.

rms pulse duration. *See* root-mean-square pulse duration.

RO. *Abbreviation for* receive only.

ROM. *Acronym for* read-only memory.

roofing filter. A low-pass filter used to reduce unwanted higher frequencies.

room noise level. *Synonym* ambient noise level.

root. In computer science, the highest level of a hierarchy.

*****root-mean-square (rms) deviation.** A single quantity, $ó_{rms}$, characterizing a function, *f*(*x*), given by

Note: The term *"rms deviation"* is also used in probability and statistics, where the normalization, M_0, is unity. Here, the term is used in a more general sense.

root-mean-square (rms) pulse broadening. The temporal rms deviation of the impulse response of a system.

root-mean-square (rms) pulse duration. A special case of root-mean-square deviation where the independent variable is time and *f*(*t*) describes the pulse waveform.

rotary dial. A signaling mechanism—usually incorporated within a telephone set—that when rotated and released, generates dc pulses required for establishing a connection in a telephone system.

rotary hunting. Hunting in which all the numbers in the hunt group are selected in a prescribed order. *Note:* In modern electronic switching systems, the numbers in the hunt group are not necessarily selected in consecutive order.

rotary switching. In telephone systems, an electro-mechanical switching method whereby the selecting mechanism consists of a rotating element using several groups of wipers, brushes, and contacts.

rotational position sensing. [In magnetic media,] A technique used to locate a given sector, a desired track, and a specific record by continuous comparison of the read/write head position with appropriate synchronization signals.

round-trip delay time. 1. The elapsed time for transit of a signal over a closed circuit. *Note:* Round-trip delay time is significant in systems that require two-way interactive communication such as voice telephony or ACK/NAK data systems where the round-trip time directly affects the throughput rate. It may range from a very few microseconds for a short line-of-sight (LOS) radio system to many seconds for a multiple-link circuit with one or more satellite links involved. This includes the node delays as well as the media transit time. 2. In primary or secondary radar systems, the time required for a transmitted pulse to reach a target and for the echo or transponder reply to return to the receiver.

route. 1. In communications systems operations, the geographical path that is followed by a call or message over the circuits that are used in establishing a chain of connections. 2. To determine the path that a message or call is to take in a communications network. *Note:* In a Transmission Control Protocol/ Internet Protocol (TCP/IP) internet, each IP datagram is routed separately. The route a datagram follows may include many gateways and many physical networks. 3. To construct the path that a call or message is to take in a communications network in going from one station to another or from a source user end instrument to a destination user end instrument.

route diversity. The allocation of circuits between two points over more than one geographic or physical route with no geographic points in common.

route matrix. In communications network operations, a record that indicates the interconnections between pairs of nodes in the network, and is used to produce direct routes, alternate routes, and available route tables from point to point.

router. In data communications, a functional unit used to interconnect two or more networks. *Note 1:* Routers operate at the network layer (layer 3) of the ISO Open Systems Interconnection–Reference Model. *Note 2:* The router reads the network layer address of all packets transmitted by a network, and forwards only those addressed to another network.

routine. A computer program, called by another program, that may have some general or frequent use.

routine message. A category of precedence to be used for all types of messages that justify transmission by rapid means unless of sufficient urgency to require a higher precedence.

routing. The process of determining and prescribing the path or method to be used for establishing telephone connections or forwarding messages.

routing diagram. In a communications system, a diagram that (a) shows all links between all switchboards, exchanges, switching centers, and stations in the system, such as the links between primary relay, major relay, minor relay, and tributary stations as well as supplementary links, (b) is used to identify the stations and links, and (c) is used to indicate tape-relay routes, transfer circuits, refile circuits, radio links, operational status, line conditions, and other network information required for network operations and management.

routing directory. *See* routing table.

routing indicator (RI). 1. A group of letters assigned to indicate: (a) the geographic location of a station; (b) a fixed headquarters of a command, activity, or unit at a geographic location; and (c) the general location of a tape relay or tributary station to facilitate the routing of traffic over the tape relay networks. 2. In a message header, an address, *i.e.,* group of characters, that specify routing instructions for the transmission of the message to its final destination. *Note:* Routing indicators may also include addresses of intermediate points.

routing table. A matrix associated with a network control protocol, which gives the hierarchy of link routing at each node.

RQ. *Abbreviation for* repeat-request. *See* ARQ.

RSL. *Abbreviation for* received signal level.

RTA. *Abbreviation for* remote trunk arrangement.

RTTY. *Abbreviation for* radio teletypewriter.

rubidium clock. A clock containing a quartz oscillator stabilized by a rubidium standard.

rubidium standard. A frequency standard in which a specified hyperfine transition of electrons in rubidium-87 atoms is used to control the output frequency. *Note:* A rubidium standard consists of a gas cell, which has an inherent long-term instability. This instability relegates the rubidium standard to its status as a secondary standard.

run. The execution of one or more computer jobs or programs.

run-length encoding. A redundancy-reduction technique for facsimile in which a run of consecutive picture elements having the same state (gray scale or color) is encoded into a single code word.

rural radio service. A public radio service rendered by fixed stations on frequencies below 1000 MHz used to provide (1) Basic Exchange Telecommunications Radio Service, which is public message communication service between a central office and subscribers located in rural areas, (2) public message communication service between landline central offices and different exchange areas which it is impracticable to interconnect by any other means, or (3) private line telephone, telegraph, or facsimile service between two or more points to which it is impracticable to extend service via landline.

rural subscriber station. A fixed station in the rural radio service used by a subscriber for communication within a central office station.

RWI. *Abbreviation for* radio and wire integration.

RX. *Abbreviation for* receive, receiver.

RZ. *Abbreviation for* return-to-zero.

s. *Abbreviation for* second.

safety service. Any radiocommunication service used permanently or temporarily for the safeguarding of human life and property.

sampling. *See* signal sampling.

sampling frequency. *See* sampling rate.

sampling interval. The reciprocal of the sampling rate, *i.e.*, the interval between corresponding points on two successive sampling pulses of the sampling signal.

sampling rate. The number of samples taken per unit time, *i.e.,* the rate at which signals are sampled for subsequent use, such as for modulation, coding, and quantization. *Deprecated synonym* sampling frequency.

sampling theorem. *Synonym* Nyquist's theorem.

satellite. A body which revolves around another body of preponderant mass and which has a motion primarily and permanently determined by the force of attraction of that other body. *Note:* A parent body and its satellite revolve about their common center of gravity.

satellite access. In satellite communications systems, the establishment of contact with a communications satellite space station. *Note:* An example of satellite access is access at the moment at which an Earth station commences to use a satellite space station as a signal repeater, *i.e.*, to use its transponder. Each radio frequency (rf) carrier that is relayed by a satellite space station at any time occupies an access channel. Accesses, *i.e.,* channels, are distinguishable by various system parameters, such as frequency, time, or code.

satellite communications. A telecommunications service provided via one or more satellite relays and their associated uplinks and downlinks.

satellite Earth terminal. *Synonym* Earth terminal.

satellite emergency position-indicating radiobeacon. An Earth station in the mobile-satellite service the emissions of which are intended to facilitate search and rescue operations.

satellite link. A radio link between a transmitting Earth station and a receiving Earth station through one satellite. A satellite link comprises one uplink and one downlink.

satellite network. A satellite system or a part of a satellite system, consisting of only one satellite and the cooperating Earth stations.

satellite operation. *See* satellite PBX.

satellite PBX. A PBX system that is not equipped with attendant positions, and is associated with an attended main PBX system. *Note:* The main attendant provides attendant functions for the satellite system.

satellite period. *See* period (of a satellite).

satellite relay. An active or passive satellite repeater that relays signals between two Earth terminals.

satellite system. A space system using one or more artificial Earth satellites.

saturation. 1. In a communications system, the condition in which a component of the system has reached its maximum traffic handling capacity. *Note:* Saturation is equivalent to one erlang per circuit. 2. The point at which the output of a linear device, such as a linear amplifier, deviates significantly from being a linear function of the input when the input signal is increased. *Note:* Modulation often requires that amplifiers operate below saturation.

scan. 1. To examine sequentially, part by part. 2. To examine every reference in every entry in a file routinely as part of a retrieval scheme. 3. In radar, one complete rotation of the interrogating antenna. 4. In SONAR, to search 360° or a specific search sector by the use of phased array of transducers. 5. To sweep, *i.e.,* rotate, a beam about a point or about an axis.

scan line. 1. The line produced on a recording medium frame by a single sweep of a scanner. 2. *Synonym* scanning line.

scanner. A device that examines a spatial pattern, one part after another, and generates analog or digital signals corresponding to the pattern. *Note:* Scanners are often used in mark sensing, pattern recognition, and character recognition.

scanning. 1. In telecommunications systems, examination of traffic activity to determine whether further processing is required. *Note:* Scanning is usually performed periodically. 2. In television, facsimile, and picture transmission, the process of successively analyzing the colors and densities of the object according to a predetermined pattern. 3. The process of tuning a device through a predetermined range of frequencies in prescribed increments and at prescribed times. *Note:* Scanning may be performed at regular or random increments and intervals. 4. In radar and radio direction finding, the slewing of an antenna or radiation pattern for the purpose of probing in a different direction. *Note 1:* In radar, scanning may be mechanical, using a rotary microwave joint to feed the antenna, or electronic, using a phased array of radiators, the radiated pattern (beam) of which depends on the relative phases of the signals fed to the individual radiators. *Note 2:* In civilian air traffic control radar, scanning usually implies continuous rotation of the antenna or beam about a vertical axis. In military radars, scanning may occur about other than a vertical axis, and may not encompass a full 360°.

scanning direction. In facsimile transmitting equipment, the scanning of an object, such as a message surface or the developed plane in the case of a drum, along parallel lines in a specified pattern. *Note 1:* The scanning direction is equivalent to scanning over a right-hand helix on a drum. *Note 2:* The orientation of the message on the scanning plane will depend upon its dimensions. *Note 3:* In facsimile receiving equipment, scanning from right to left and top to bottom, is called "positive" reception and from left to right and top to bottom, is called "negative" reception. *Note 4:* Scanning direction conventions are included in CCITT Recommendations for phototelegraphic equipment.

scanning field. In facsimile systems, the total of the areas that are actually explored by the scanning spot during the scanning of the object by the transmitter or during scanning of the record medium by the receiver.

scanning line. In an imaging system, the path traversed by a scanning spot during a single line sweep.

scanning line frequency. In facsimile, the frequency at which a fixed line perpendicular to the direction of scanning is crossed by a scanning spot. *Note:* The scanning line frequency is equivalent to drum speed in some mechanical systems. *Synonym* scanning line rate.

scanning line length. In facsimile systems, the total length of a scanning line, equal to the spot speed divided by the scanning line frequency. *Note:* The scanning line length is usually greater than the length of the available line.

scanning line period. In facsimile systems, the time interval between (a) the instant at which the scanning spot probes or writes to a given spot on one scanning line, and (b) the instant at which the scanning spot probes or writes to the corresponding spot on the next scanning line.

scanning line rate. *Synonym* scanning line frequency.

scanning pitch. The distance between the centers of consecutive scanning lines.

scanning rate. In facsimile and television systems, the rate of displacement of the scanning spot along the scanning line.

scanning spot. In facsimile systems, the area on the object, *i.e.,* the original, covered instantaneously by the pickup system of the scanner.

scan-stop lockup. In automatic link establishment (ALE) radios, the undesired condition in which the normal process of (a) scanning radio channels, (b) stopping on the desired channel, or (c) returning to scan is terminated by the equipment.

scatter. *See* scattering.

scattering. Of a wave propagating in a material medium, a phenomenon in which the direction, frequency, or polarization of the wave is changed when the wave encounters discontinuities in the medium, or interacts with the material at the atomic or molecular level. *Note:* Scattering results in a disordered or random change in the incident energy distribution.

scattering center. In the microstructure of a transmission medium, a site at which electromagnetic waves are scattered. *Note 1:* Examples of scattering centers are vacancy defects; interstitial defects; inclusions, such as a gas molecules, hydroxide ions, iron ions, and trapped water molecules; and microcracks or fractures in dielectric waveguides. *Note 2:* Scattering centers are frozen in the medium when it solidifies and may not necessarily cause Rayleigh scattering, which varies inversely as the fourth power of the wavelength. For example, in glass optical fibers, there is a high attenuation band at 0.95 μm, primarily caused by scattering and absorption by OH^- (hydroxyl) ions.

scattering coefficient. The factor that expresses the attenuation caused by scattering, *e.g.*, of radiant or acoustic energy, during its passage through a medium. *Note:* The scattering coefficient is usually expressed in units of reciprocal distance.

scattering cross section. The area of an incident wavefront, at a reflecting surface or medium, such as an object in space, through which will pass radiant energy, that, if isotropically scattered from that point, would produce the same power at a given receiver as is actually provided by the entire reflecting surface.

scattering loss. The part of the transmission (power) loss that results from scattering within a transmission medium or from roughness of a reflecting surface.

SCC. *Abbreviation for* specialized common carrier.

scene cut. Video imagery in which consecutive frames are highly uncorrelated.

scene cut response. In video systems, the perceived impairments associated with a scene cut.

schematic. 1. A diagram, drawing, or sketch that details the elements of a system, such as the elements of an electrical circuit or the elements of a logic diagram for a computer or communications system. 2. Pertaining to a diagram, drawing, or sketch that details the elements of a system, such as the elements of an electrical circuit or the elements of a logic diagram for a computer or communications system.

scintillation. In electromagnetic wave propagation, a small random fluctuation of the received field strength about its mean value. *Note:* Scintillation effects become more significant as the frequency of the propagating wave increases.

scrambler. A device that transposes or inverts signals or otherwise encodes a message at the transmitter to make the message unintelligible at a receiver not equipped with an appropriately set descrambling device. *Note:* Scramblers usually use a fixed algorithm or mechanism. However, a scrambler provides communications privacy that is inadequate for classified traffic.

screen. 1. In a telecommunications, computing, or data processing system, to examine entities that are being processed to determine their suitability for further processing. 2. A nonferrous metallic mesh used to provide electromagnetic shielding. 3. To reduce undesired electromagnetic signals and noise by enclosing devices in electrostatic or electromagnetic shields. 4. A viewing surface, such as that of a cathode ray tube or liquid crystal display (LCD).

scroll. In a display device, to move the display window of the screen vertically to view the contents of a stored document. *Note:* Scrolling may be performed continuously or incrementally.

SDLC. *Abbreviation for* synchronous data link control.

search time. In data processing systems, the time interval required to locate a particular data element, record, or file in a storage device.

SECAM. *Acronym for* système electronique couleur avec memoire. A television signal standard (625 lines, 50 Hz, 220 V primary power) used in France, eastern European countries, the former USSR, and some African countries.

second (s). In the International System of Units (SI), the time interval equal to 9,192,631,770 periods of the radiation corresponding to the transition between the two hyperfine levels of the ground state of the cesium-133 atom.

secondary channel. In a system in which two channels share a common interface, a channel that has a lower data signaling rate (DSR) capacity than the primary channel.

secondary emission. Particles or radiation, such as photons, Compton recoil electrons, delta rays, secondary cosmic rays, and secondary electrons, that are produced by the action of primary radiation on matter.

secondary frequency standard. A frequency standard that does not have inherent accuracy, and therefore must be calibrated against a primary frequency standard. *Note:* Secondary standards include crystal oscillators and rubidium standards. A crystal oscillator depends for its frequency on its physical dimensions, which vary with fabrication and environmental conditions. A rubidium standard is a secondary standard even though it uses atomic transitions, because it takes the form of a gas cell through which an optical signal is passed. The gas cell has inherent inaccuracies because of gas pressure variations, including those induced by temperature variations. There are also variations in the concentrations of the required buffer gases, which variations cause frequency deviations.

secondary radar. A radiodetermination system based on the comparison of reference signals with radio signals retransmitted from the position to be determined. *Note:* An example of secondary radar is the transponder-based surveillance of aircraft. *Synonym* secondary surveillance radar.

secondary radiation. *See* secondary emission.

secondary service area. [T]he service area of a broadcast station served by the skywave and not subject to objectionable interference and in which the signal is subject to intermittent variations in strength.

secondary station. In a communications network, a station that (a) is responsible for performing unbalanced link-level operations as instructed by the primary station and (b) interprets received commands and generates responses.

secondary surveillance radar. *Synonym* secondary radar.

secondary time standard. A time standard that requires periodic calibration against a primary time standard.

second dialtone. 1. Dialtone presented to the call originator after an access code has been dialed for access to a second, outside, telecommunications system or service. 2. Dialtone returned to the call originator after she/he has dialed an access number and has reached a switch providing access to modem, to a fax machine, to another telephone, *etc.*

second window. Of silica-based optical fibers, the transmission window at approximately 1.3 μm. *Note:* The second window is the minimum-dispersion window in silica-based glasses.

SECORD. *Acronym for* secure voice cord board. A desk-mounted patch panel that provides the capability for controlling (a) sixteen 50-kb/s wideband or sixteen 2400-b/s narrowband user lines and (b) 5 narrowband trunks to DSN or other narrowband facilities.

SECTEL. *Acronym for* secure telephone. *See* STU.

sector. A predetermined, addressable angular part of a track or band on a magnetic drum or magnetic disk.

sectoring. In magnetic or optical disk storage media, the division of tracks into a specified number of segments, for the purpose of organizing the data stored thereon.

secure communications. Telecommunications deriving security through use of type 1 products and/or protected distribution systems.

secure telephone unit. *See* STU.

secure transmission. 1. In transmission security, *see* secure communications. 2. In spread-spectrum systems, the transmission of binary coded sequences that represent information that can be recovered only by persons or systems that have the proper key for the spread-spectrum code-sequence generator, *i.e.,* have a synchronized generator that is identical to that used for transmission.

secure voice cord board. *See* SECORD.

security. 1. A condition that results from the establishment and maintenance of protective measures that ensure a state of inviolability from hostile acts or influences. 2. With respect to classified matter, the condition that prevents unauthorized persons from having access to official information that is safeguarded in the interests of national security. 3. Measures taken by a military unit, an activity or installation to protect itself against all acts designed to, or which may, impair its effectiveness.

security filter. 1. In communications security, the hardware, firmware, or software used to prevent access to specified data by unauthorized persons or systems, such as by preventing transmission, preventing forwarding messages over unprotected lines or circuits, or requiring special codes for access to read-only files. 2. [An] AIS trusted subsystem that enforces security policy on the data that passes through it.

security kernel. 1. In computer and communications security, the central part of a computer or communications system hardware, firmware, and software that implements the basic security procedures for controlling access to system resources. 2. A self-contained usually small collection of key security-related statements that (a) works as a part of an operating system to prevent unauthorized access to, or use of, the system and (b) contains criteria that must be met before specified programs can be accessed. 3. Hardware, firmware, and software elements of a trusted computing base that implement the reference monitor concept.

security management. In network management, the set of functions (a) that protects telecommunications networks and systems from unauthorized access by persons, acts, or influences and (b) that includes many subfunctions, such as creating, deleting, and controlling security services and mechanisms; distributing security-relevant information; reporting security-relevant events; controlling the distribution of cryptographic keying material; and authorizing subscriber access, rights, and privileges.

seek. To position selectively the access mechanism of a direct access [storage] device.

seek time. The time required for the access arm of a direct-access storage device to be positioned on the appropriate track. *Synonym* positioning time.

segment. In a distributed queue dual bus (DQDB) network, a protocol data unit (PDU) that (a) consists of 52 octets transferred between DQDB-layer peer entities as the information payload of a slot, (b) contains a header of 4 octets and a payload of 48 octets, and (c) is either a pre-arbitrated segment or a queued arbitrated segment.

segmented encoding law. An encoding law in which an approximation to a curve defined by a smooth encoding law is obtained by a number of linear segments. *Synonym* piecewise linear encoding law.

segregation. *Synonym* privacy (def. #1).

seizing. The temporary dedication of various parts of a communications system to a specific use, usually in response to a user request for service. *Note:* The parts seized may be automatically connected, such as by direct distance dialing (DDD), or may require operator intervention.

seizure signal. In telephone systems, a signal used by the calling end of a trunk or line to indicate a request for service. *Note:* A seizure signal also locks out the trunk or line to other demands for service.

selcall. *Acronym for* selective calling. Calling from one station in which call identification is sent to signal automatically one or more remote stations and to establish links among them. *Note 1:* Selective calling may be used to un-mute the speakers at designated stations or to initiate a handshake for link establishment. *Note 2:* Selective calling is specified in

CCIR Recommendations for HF and VHF/UHF radio, generally for ship-to-shore, ship-to-ship, aircraft-to-aircraft, and aircraft-to-ground communications.

selection position. *Synonym* decision instant.

selective calling. *See* selcall.

selective combiner. *Synonym* maximal-ratio combiner.

selective fading. Fading in which the components of the received radio signal fluctuate independently.

selective jamming. *See* electronic warfare.

selective ringing. In a party line, ringing only the desired user instrument. *Note:* Without selective ringing, all the instruments on the party line will ring at the same time, selection being made by the number of rings.

selectivity. A measure of the ability of a receiver to discriminate between a wanted signal on one frequency and unwanted signals on other frequencies.

self-authentication. 1. A procedure in which a transmitting station, *i.e.,* a calling station, establishes its own validity without the participation of the receiving station, *i.e.,* the called station. *Note:* The calling station establishes its own authenticity and the called station is not required to challenge the calling station. Self-authentication is usually used only when one-time authentication systems are used to derive the authentication. 2. Implicit authentication, to a predetermined level, of all transmissions on a secure communications system.

self-delineating block. A block in which a bit pattern or a flag identifies the beginning or end of a block.

self-synchronizing code. A code in which the symbol stream formed by a portion of one code word, or by the overlapped portion of any two adjacent code words, is not a valid code word. *Note 1:* A self-synchronizing code permits the proper framing of transmitted code words provided that no uncorrected errors occur in the symbol stream. *Note 2:* External synchronization is not required. *Note 3:* High-level data link control (HDLC) and Advanced Data Communication Control Procedures (ADCCP) frames represent self-synchronizing code words.

semiautomated tactical command and control system. A machine-aided command and control system in which human intervention is required in varying degrees to operate the system.

semiautomatic switching system. 1. In telephone systems, a switching system in which telephone operators receive call instructions orally from users and complete them by automatic equipment. 2. At tape-relay intermediate stations, the manual routing or rerouting of taped messages without rekeying them.

semiconductor laser. *Synonym* injection laser diode.

semiduplex operation. 1. A method which is simplex operation at one end of the circuit and duplex operation at the other. *RR Footnote:* In general, duplex operation and semiduplex operation require two frequencies in radiocommunication; simplex operation may use either one or two. 2. Operation of a communications network in which a base station operates in a duplex mode with a group of remote stations operating in a half-duplex mode. *Note:* The terms *"half-duplex"* and *"simplex"* are used differently in wire and radio communications.

sender. A device that accepts address information from a register or routing information from a translator, and then transmits the proper routing information to a trunk or to local equipment. *Note:* Sender and register functions are often combined in a single unit.

sending-end crossfire. In teletypewriter (TTY) systems, interference, in a given channel, caused by transmissions from one or more adjacent TTY channels transmitting from the end at which the crossfire, *i.e.*, interference, is measured.

sensitive information. Information, the loss, or misuse, or unauthorized access to or modification of which could adversely affect the national interest or the conduct of federal programs, or the privacy to which individuals are entitled to under 5 U.S.C. Section 552a (the Privacy Act), but that has not been specifically authorized under criteria established by an Executive Order or an Act of Congress to be kept secret in the interest of national defense or foreign policy.

sensitivity. In an electronic device, *e.g.*, a communications system receiver, or detection device, *e.g.*, PIN diode, the minimum input signal required to produce a specified output signal having a specified signal-to-noise ratio, or other specified criteria. *Note 1:* The signal input may be expressed as power in dBm or as field strength in microvolts per meter, with input network impedance stipulated. *Note 2: "Sensitivity"* is sometimes improperly used as a synonym for *"responsivity."*

sensor. A device that responds to a physical stimulus, such as thermal energy, electromagnetic energy, acoustic energy, pressure, magnetism, or motion, by producing a signal, usually electrical.

sentinel. *See* flag.

separate channel signaling. Signaling in which the whole or a part of one or more channels in a multichannel system is used to provide for supervisory and control signals for the message traffic channels. *Note:* The same channels, such as frequency bands or time slots, that are used for signaling are not used for message traffic. *Contrast with* common-channel signaling.

septet. A byte composed of seven binary elements. *Synonym* seven-bit byte.

sequence. An arrangement of items according to a specified set of rules, for example, items arranged alphabetically, numerically, or chronologically.

sequential access. *Synonym* serial access.

sequential logic element. A device that has at least one output channel and one or more input channels, all characterized by discrete states, such that the state of each output channel is determined by the previous states of the input channels.

sequential transmission. *Synonym* serial transmission.

serial. 1. Pertaining to a process in which all events occur one after the other; for example, the serial transmission of the bits of a character according to the CCITT V.25 protocol. 2. Pertaining to the sequential or consecutive occurrence of two or more related activities in a single device or channel. 3. Pertaining to the sequential processing of the individual parts of a whole, such as the bits of a character or the characters of a word, using the same facilities for successive parts.

serial access. 1. Pertaining to the sequential or consecutive transmission of data into or out of a device, such as a computer, transmission line, or storage device. 2. A process by which data are obtained from a storage device or entered into a storage device in such a way that the process depends on the location of those data and on a reference to data previously accessed. *Synonym* sequential access.

serial computer. 1. A computer that has a single arithmetic and logic unit. 2. A computer, some specified characteristic of which is serial; for example, a computer than manipulates all bits of a word serially.

serializer. *See* parallel-to-serial conversion.

serial port. A port through which data are passed serially, *i.e.*, one bit at a time, and that requires only one input channel to handle a set of bits, *e.g.*, all the bits of a byte. *Contrast with* parallel port.

serial-to-parallel conversion. Conversion of a stream of data elements received in time sequence, *i.e.*, one at a time, into a data stream consisting of multiple data elements transmitted simultaneously. *Contrast with* **paral**lel-to-serial conversion.

serial transmission. The sequential transmission of the signal elements of a group representing a character or other entity of data. *Note:* The characters are transmitted in a sequence over a single line, rather than simultaneously over two or more lines, as in parallel transmission. The sequential elements may be transmitted with or without interruption. *Synonym* sequential transmission.

series T junction. A three-port waveguide junction that has an equivalent circuit in which the impedance of the branch waveguide is predominantly in series with the impedance of the main waveguide at the junction.

server. A network device that provides service to the network users by managing shared resources. *Note 1:* The term is often used in the context of a client-server architecture for a local area network (LAN). *Note 2:* Examples are a printer server and a file server.

service. In the Open Systems Interconnection–Reference Model (OSI–RM), a capability of a given layer, and the layers below it, that (a) is provided to the entities of the next higher layer and (b) for a given layer, is provided at the interface between the given layer and the next higher layer.

service access. In personal communications service (PCS), the ability for the network to provide user access to features and to accept user service requests specifying the type of bearer services or supplementary service that the users want to receive from the PCS network.

service access point (SAP). 1. A physical point at which a circuit may be accessed. 2. In an Open Systems Interconnection (OSI) layer, a point at which a designated service may be obtained.

service bit. A system overhead bit used for providing a network service, such as a request for a repetition or for a numbering sequence. *Note:* A service bit is not a check bit.

service channel. *Synonym* orderwire circuit.

service class. *See* class of service.

service data unit (SDU). In layered systems, a set of data that is sent by a user of the services of a given layer, and is transmitted to a peer service user semantically unchanged.

service feature. In telephony, any of a number of special functions that may be specified initially, or added to, the user's basic service. *Note:* Modern telephone switches are capable of providing a wide variety of service features, such as call forwarding and call waiting.

service identification. The information that uniquely identifies an NS/EP telecommunications service to the service vendor and the service user.

service integrity. The degree to which a service is provided without excessive impairment, once obtained. [NATO]

service probability. The probability of obtaining a specified (or higher) grade of service during a given period of time.

service profile. *Synonym* UPT service profile.

service profile management. *Synonym* UPT service profile management.

service program. *Synonym* utility program.

service routine. *Synonym* utility program.

service signals. Signals that enable data systems equipment to function correctly, and possibly to provide ancillary facilities. *Synonym* housekeeping signals.

service termination point. The last point of service rendered by a commercial carrier under applicable tariffs. *Note 1:* The service termination point is usually on the customer premises. *Note 2:* The customer is responsible for equipment and operation from the service termination point to user end instruments. *Note 3:* The service termination point usually corresponds to the demarcation point.

service user. An individual or organization, including a service vendor, that is provided a telecommunications service for which a priority level has been requested or assigned.

Session Layer. *See* Open Systems Interconnection—Reference Model.

set. 1. A finite or infinite number of objects, entities, or concepts, that have a given property or properties in common. 2. To configure all or part of a device into a specified state.

seven-bit byte. *Synonym* septet.

seven-hundred (700) service. A personal telephone service that allows individuals to receive, via a single number, telephone calls in various locations (*e.g.*, home, office, or car) from call originators using the same common carrier.

sexadecimal. *Synonym* hexadecimal.

sextet. A byte composed of six binary elements. *Synonym* six-bit byte.

S-F. *Abbreviation for* store-and-forward.

SF. *Abbreviation for* single-frequency. *See* single-frequency signaling.

SFTS. *Abbreviation for* standard frequency and time signal. *See* standard time and frequency signal service.

SGML. *Abbreviation for* Standard Generalized Mark-up Language. A file format for storage of text and graphics files.

shadow loss. 1. The attenuation caused to a radio signal by obstructions in the propagation path. 2. In a reflector antenna, the relative reduction in the effective aperture of the antenna caused by the masking effect of other antenna parts, such as a feed horn or a secondary reflector, which parts obstruct the radiation path.

shannon (Sh). The unit of information derived from the occurrence of one of two equiprobable, mutually exclusive, and exhaustive events. *Note:* A bit may, with perfect formatting and source coding, contain 1 Sh of information. However, the information content of a bit is usually be less than 1 Sh.

Shannon's law. A statement defining the theoretical maximum rate at which error-free digits can be transmitted over a bandwidth-limited channel in the presence of noise, usually expressed in the form $C = W \log_2 (1 + S/N)$, where C is the channel capacity in bits per second, W is the bandwidth in hertz, and S/N is the signal-to-noise ratio. *Note:* Error-correction codes can improve the communications performance relative to uncoded transmission, but no practical error correction coding system exists that can closely approach the theoretical performance limit given by Shannon's law.

shaping network. A network inserted in a circuit for the purpose of improving or modifying the waveform of signals.

sheath. Of a communications or power cable, the outer covering or coverings of tough material, often plastic, that is resistant to environmental hazards such as abrasion, liquid intrusion, solar radiation, *etc.,* and is used to protect cable component(s) such as optical fibers or metallic conductors that transport the signal or power. *Note:* There may be more than one sheath surrounding a given cable. For example, some cable designs use an inner sheath surrounded by metallic armor, over which is an outer sheath. *Synonym* jacket.

sheath miles. The actual length of cable in route miles.

shell. In a computer environment, an operating system command interpreter, *i.e.,* a software utility that reads an input specifying an operation, and that may perform, direct, or control the specified operation. *Note 1:* For example, a shell may permit a user to switch among application programs without terminating any of them. *Note 2:* A shell may take its input from either a user terminal or from a file.

SHF. *Abbreviation for* super high frequency. *See* electromagnetic spectrum.

shield. 1. A housing, screen, sheath, or cover that substantially reduces the coupling of electric, magnetic, or electromagnetic fields into or out of circuits or transmission lines. 2. A protective cover that prevents the accidental contact of objects or persons with parts or components operating at hazardous voltage levels.

shielded pair. A 2-wire transmission line surrounded by a sheath of conductive material that protects it from the effects of external fields and confines fields produced within the line.

shielded twisted pair. A transmission line composed of a twisted 2-wire metallic transmission line surrounded by a sheath of conductive material that protects it from the effects of external fields and confines fields produced within the line.

shielding. *See* shield.

shielding effectiveness. The factor that expresses the attenuation caused by scattering, *e.g.*, of radiant or acoustic energy, during its passage through a medium. *Note:* The scattering coefficient is usually expressed in units of reciprocal distance.

shift. 1. The movement of some or all of the characters or bits of a word by the same number of character or bit positions in the direction of a specified end of a word. 2. In radar, the ability to move the origin of a radial display away from the center of the cathode ray tube.

shift register. A storage device, usually in a central processing unit (CPU), in which device a serially ordered set of data may be moved, as a unit, into a discrete number of storage locations. *Note 1:* Shift registers may be configured so that the stored data may be moved in more than one direction. *Note 2:* Shift registers may be configured so that data may be entered and stored from multiple inputs. *Note 3:* Shift registers may be grouped into arrays of two or more dimensions in order to perform more complex data operations.

ship Earth station. A mobile Earth station in the maritime mobile-satellite service located on board ship.

ship's emergency transmitter. A ship's transmitter to be used exclusively on a distress frequency for distress, urgency or safety purposes.

ship station. A mobile station in the maritime mobile service located on board a vessel which is not permanently moored, other than a survival craft station.

shock excitation. *Synonym* impulse excitation.

short haul toll traffic. A general term applied to message toll traffic between nearby points. In common usage, this term is ordinarily applied to message toll traffic between points less than 20 to 50 miles apart.

shortwave. In radio communications, pertaining to the band of frequencies approximately between 3 MHz and 30 MHz. *Note: "Shortwave"* is not a term officially recognized by the international community.

short wavelength. In optical communication, optical radiation having a wavelength less than approximately 1 μm.

shot noise. The noise caused by random fluctuations in the motion of charge carriers in a conductor. *Note:* There is often a minor inconsistency in referring to shot noise in an optical system: many authors refer to shot noise loosely when speaking of the mean square shot noise current (amperes^2) rather than noise power (watts).

SI. *Abbreviation for* International System of Units.

SID. *Abbreviation for* sudden ionospheric disturbance.

sideband. In amplitude modulation (AM), a band of frequencies higher than or lower than the carrier frequency, containing energy as a result of the modulation process. *Note:* Amplitude modulation results in two sidebands. The frequencies above the carrier frequency constitute what is referred to as the *"upper sideband"*; those below the carrier frequency, constitute the *"lower sideband."* In conventional AM transmission, both sidebands are present. Transmission in which one sideband is removed is called *"single-sideband transmission."*

sideband transmission. *See* single-sideband transmission.

side circuit. Either of the two circuits used to derive a phantom circuit.

side lobe. In a directional antenna radiation pattern, a lobe in any direction other than that of the main lobe.

sidetone. The sound of the speaker's own voice (and background noise) as heard in the speaker's telephone receiver. *Note:* Sidetone volume is usually suppressed relative to the transmitted volume. *Synonym* telephone sidetone.

SIGINT. *Acronym for* signals intelligence.

signal. 1. Detectable transmitted energy that can be used to carry information. 2. A time-dependent variation of a characteristic of a physical phenomenon, used to convey information. 3. As applied to electronics, any transmitted electrical impulse. 4. Operationally, a type of message, the text of which consists of one or more letters, words, characters, signal flags, visual displays, or special sounds, with prearranged meaning and which is conveyed or transmitted by visual, acoustical, or electrical means.

signal center. A combination of signal communication facilities operated by the Army in the field and consisting of a communications center, telephone switching central and appropriate means of signal communications.

signal compression. 1. In analog (usually audio) systems, reduction of the dynamic range of a signal by controlling it as a function of the inverse relationship of its instantaneous value relative to a specified reference level. *Note 1:* Signal compression is usually expressed in dB. *Note 2:* Instantaneous values of the input signal that are low, relative to the reference level, are increased, and those that are high are decreased. *Note 3:* Signal compression is usually accomplished by separate devices called *"compressors."* It is used for many purposes, such as (a) improving signal-to-noise ratios prior to digitizing an analog signal for transmission over a digital carrier system, (b) preventing overload of succeeding elements of a system, or (c) matching the dynamic ranges of two devices. *Note 4:* Signal compression (in dB) may be a linear or nonlinear function of the signal level across the frequency band of interest and may be essentially instantaneous or have fixed or variable delay times. *Note 5:* Signal compression always introduces distortion, which is usually not objectionable, if the compression is limited to a few dB. *Note 6:* The original dynamic range of a compressed signal may be restored by a circuit called an *"expander."* 2. In facsimile systems, a process in which the number of pels scanned on the original is larger than the number of encoded bits of picture information transmitted.

signal contrast. In facsimile, the ratio of the level of the white signal to the level of the black signal. *Note:* Signal contrast is usually expressed in dB.

signal conversion equipment. *Synonym* modem.

signal distance. 1. A measure of the difference between a given signal and a reference signal. *Note:* For analog signals, the signal distance is the root mean square difference between the given signal and a reference signal over a symbol period. 2. *Synonym* Hamming distance.

signal distortion. *See* distortion.

signal droop. In an otherwise essentially flat-topped rectangular pulse, distortion characterized by a decline of the pulse top. *See illustration under* waveform.

signal element. A part of a signal, distinguished by its nature, magnitude, duration, transition, or relative position. *Note:* Examples of signal elements include signal transitions, significant conditions, significant instants, and binary digits (bits).

signal expansion. Restoration of the dynamic range of a compressed signal. *Contrast with* **signal compression.**

signal frequency shift. *See* frequency shift.

signaling. 1. The use of signals for controlling communications. 2. In a telecommunications network, the information exchange concerning the establishment and control of a connection and the management of the network, in contrast to user information transfer. 3. The sending of a signal from the transmitting end of a circuit to inform a user at the receiving end that a message is to be sent.

signaling path. In a transmission system, a path used for system control, synchronization, checking, signaling, and service signals used in system management and operations rather than for the data, messages, or calls of the users.

signaling rate. *See* data signaling rate.

Signaling System No. 7 (SS7). A common-channel signaling system defined by the CCITT in the 1988 Blue Book, in Recommendations Q.771 through Q.774. *Note:* SS7 is a prerequisite for implementation of an Integrated Services Digital Network (ISDN).

signaling time slot. In TDM carrier systems, a time slot starting at a particular phase or instant in each frame and allocated to the transmission of signaling (supervisory and control) data.

signal intelligence. *See* signals intelligence.

signal level. In a communications system, the signal power or intensity at a specified point and with respect to a specified reference level, *e.g.*, 1 mW.

signal message. In communications systems, a message, *i.e.*, an assembly of signaling information, that (a) includes associated message alignment and service indications, (b) pertains to a call, and (c) is transferred via the message transfer part.

signal-plus-noise-plus-distortion to noise-plus-distortion ratio. *See* SINAD.

signal-plus-noise-to-noise ratio ((*S+N*)/*N*). At a given point in a communications system, the ratio of (a) the power of the desired signal plus the noise to (b) the power of the noise. *Note:* The (*S+N*)/*N* ratio is usually expressed in dB.

signal processing. The processing–such as detection, shaping, converting, coding, and time positioning–of signals, that results in their transformation into other forms, such as other waveshapes, power levels, and coding arrangements.

signal processing gain. 1. The ratio of (a) the signal-to-noise ratio of a processed signal to (b) the signal-to-noise ratio of the unprocessed signal. *Note:* Signal processing gain is usually expressed in dB. 2. In a spread-spectrum communications system, the signal gain, signal-to-noise ratio, signal shape, or other signal improvement obtained by coherent band spreading, remapping, and reconstitution of the desired signal.

signal reference subsystem. The portion of a facility grounding system that (a) provides reference planes, such as ground-return circuits, for all of the signal paths in the facility and (b) is isolated from other circuits, especially isolated from circuits that carry fault, lightning discharge, and power distribution currents.

signal regeneration. Signal processing that restores a signal so that it conforms to its original characteristics.

signal-return circuit. A current-carrying return path from a load back to the signal source, *i.e.,* the low side of the closed loop energy transfer circuit between a source-load pair.

signal sample. The value of a particular characteristic of a signal at a chosen instant.

signal sampling. The process of obtaining a sequence of instantaneous values of a particular signal characteristic, usually at regular time intervals.

signal security. A generic term that includes both communications security and electronics security.

signals intelligence (SIGINT). 1. A category of intelligence comprising, either individually or in combination, all communications intelligence, electronics intelligence, and foreign instrumentation signals intelligence, however transmitted. 2. Intelligence derived from communications, electronics, and foreign instrumentation signals.

signals security. [A] generic term encompassing communications security and electronic security.

signal-to-crosstalk ratio. At a specified point in a circuit, the ratio of the power of the wanted signal to the power of the unwanted signal from another channel. *Note 1:* The signals are adjusted in each channel so that they are of equal power at the zero transmission level point in their respective channels. *Note 2:* The signal-to-crosstalk ratio is usually expressed in dB.

signal-to-noise ratio (SNR). The ratio of the amplitude of the desired signal to the amplitude of noise signals at a given point in time. *Note 1:* SNR is expressed as 20 times the logarithm of the amplitude ratio, or 10 times the logarithm of the power ratio. *Note 2:* SNR is usually expressed in dB and in terms of peak values for impulse noise and root-mean-square values for random noise. In defining or specifying the SNR, both the signal and noise should be characterized, *e.g.,* peak-signal-to-peak-noise ratio, in order to avoid ambiguity.

signal-to-noise ratio per bit. The ratio given by E_b/N_0, where E_b is the signal energy per bit and N_0 is the noise energy per hertz of noise bandwidth.

signal transfer point (STP). In a common-channel signaling network, a switching center that provides for the transfer from one signaling link to another. *Note:* In nonassociated common-channel signaling, the signal transfer point need not be the point through which the call, which is associated with the signaling being switched, passes.

signal transition. In the modulation of a carrier, a change from one significant condition to another. *Note 1:* Examples of signal transitions are a change from one electrical current, voltage, or power level to another; a change from one optical power level to another; a phase shift; or a change from one frequency or wavelength to another. *Note 2:* Signal transitions are used to create signals that represent information, such as "0" and "1" or "mark" and "space."

signature. 1. The complete set of electromagnetic and/or acoustic signals received, *e.g.,* from an infrared source, a radio or radar transmitter, an aircraft, or a ship. *Note:* Signatures may consist of analog or digital signals, or both, and may be analyzed to indicate the nature of their source and assist in its recognition. 2. The attributes of an electromagnetic or acoustic wave that has been reflected from or transmitted through an object and contains information indicating the attributes of the object.

significant condition. In the modulation of a carrier, one of the values of the signal parameter chosen to represent information. *Note 1:* Examples of significant conditions are an electrical current, voltage, or power level; an optical power level; a phase value; or a frequency or wavelength chosen to represent a "0" or a "1"; or a "mark" or a "space." *Note 2:* The duration of a significant condition is the time interval between successive significant instants. *Note 3:* A change from one significant condition to another is called a *"signal transition." Note 4:* Signal transitions are used to create signals that represent information, such as "0" and "1" or "mark" and "space." *Note 5:* Significant conditions are recognized by an appropriate device. Each significant instant is determined when the appropriate device assumes a condition or state usable for performing a specific function, such as recording, processing, or gating.

significant digit. In a representation of a number, a digit that is needed for a given purpose; in particular, a digit that must be kept to preserve a given accuracy or a given precision.

significant instant. In a signal, any instant at which a significant condition of a signal begins or ends. *Note:* Examples of significant instants include the instant at which a signal crosses the baseline or reaches 10% or 90% of its maximum value.

significant interval. The time interval between two consecutive significant instants.

silent zone. *Synonym* skip zone.

silica. Silicon dioxide (SiO_2). *Note 1:* Silica may occur in crystalline or amorphous form, and occurs naturally in impure forms such as quartz and sand. *Note 2:* Silica is the basic material of which the most common communication-grade optical fibers are presently made.

silicon dioxide (SiO_2). *See* silica.

silicon photodiode. A silicon-based PN- or PIN-junction photodiode. *Note 1:* Such photodiodes are useful for direct detection of optical wavelengths shorter than approximately 1 µm. *Note 2:* Because of their greater bandgap, silicon-based photodiodes are quieter than germanium-based photodiodes, but germanium photodiodes must be used for wavelengths longer than approximately 1 µm.

simple buffering. The assigning of buffer storage for the duration of the execution of a computer program.

Simple Mail Transfer Protocol (SMTP). The Transmission Control Protocol/Internet Protocol (TCP/IP) standard protocol that facilitates transfer of electronic-mail messages, specifies how two systems are to interact, and specifies the format of messages used to control the transfer of electronic mail.

Simple Network Management Protocol (SNMP). The Transmission Control Protocol/Internet Protocol (TCP/IP) standard protocol that (a) is used to manage and control IP gateways and the networks to which they are attached, (b) uses IP directly, bypassing the masking effects of TCP error correction, (c) has direct access to IP datagrams on a network that may be operating abnormally, thus requiring management, (d) defines a set of variables that the gateway must store, and (e) specifies that all control operations on the gateway are a side-effect of fetching or storing those data variables, *i.e.,* operations that are analogous to writing commands and reading status.

simple scanning. In facsimile transmission, scanning using only one spot at a time.

simplex circuit. 1. A circuit that provides transmission in one direction only. 2. *Deprecated definition:* A circuit using ground return and affording communication in either direction, but in only one direction at a time. *Note:* The above two definitions are contradictory; however, both are in common use. The user is cautioned to verify the nature of the service specified by this term.

simplex operation. 1. Operation in which transmission occurs in one and only one preassigned direction. *Synonym* one-way operation. *Note:* Duplex operation may be achieved by simplex operation of two or more simplex circuits or channels. 2. Operating method in which transmission is made possible alternately in each direction of a telecommunication channel, for example by means of manual control. *Note:* In general, duplex operation and semiduplex operation require two frequencies in radiocommunication; simplex operation may use either one or two. *Note 2:* These two definitions are contradictory, however, both are in common use. The first one is used in telephony and the second one is used in radio. The user is cautioned to verify the nature of the service specified by this term.

simplex (SX) signaling. Signaling in which two conductors are used for a single channel, and a center-tapped coil, or its equivalent, is used to split the signaling current equally between the two conductors. *Note:* SX signaling may be one-way, for intra-central-office use, or the simplex legs may be connected to form full duplex signaling circuits that function like composite (CX) signaling circuits with E & M lead control.

simulate. To represent certain features of the behavior of a physical or abstract system by the behavior of another system. *Note 1:* For example, delay lines may be used to simulate propagation delay and phase shift caused by an actual transmission line. *Note 2:* A simulator may imitate only a few of the operations and functions of the unit it simulates. *Contrast with* emulate.

SINAD. *Abbreviation for* signal-plus-noise-plus-distortion to noise-plus-distortion ratio. 1. The ratio of (a) total received power, *i.e.*, the received signal-plus-noise-plus-distortion power to (b) the received noise-plus-distortion power. 2. The ratio of (a) the recovered audio power, *i.e.,* the original modulating audio signal plus noise plus distortion powers from a modulated radio frequency carrier to (b) the residual audio power, *i.e.,* noise-plus-distortion powers remaining after the original modulating audio signal is removed. *Note:* The SINAD is usually expressed in dB.

singing. An undesired self-sustained audio oscillation in a circuit. *Note:* Singing is usually caused by positive feedback, excessive gain, or unbalance of a hybrid termination, or by some combination of these.

singing margin. The difference in power levels between the singing point and the operating gain of a system or component.

singing point. The threshold point at which additional gain in the system will cause self-oscillation.

single-current system. *Synonym* neutral direct-current telegraph system.

single-current transmission system. *Synonym* neutral direct-current telegraph system.

single-ended control. *Synonym* single-ended synchronization.

single-ended synchronization. Synchronization between two locations, in which phase error signals used to control the clock at one location are derived by comparing the phase of the incoming signals to the phase of the internal clock at that location. *Synonym* single-ended control.

single-frequency interference. Interference caused by a single-frequency source. *Note 1:* An example of single-frequency interference is interference in a transmission channel induced by a 60-Hz source. *Note 2:* The interference caused by the single-frequency source may have other frequencies and may also appear in many channels.

single-frequency (SF) signaling. In telephony, signaling in which dial pulses or supervisory signals are conveyed by a single voice-frequency tone in each direction. *Note 1:* An SF signaling unit converts E & M signaling to a format (characterized by the presence or absence of a single voice-frequency tone), which is suitable for transmission over an ac path, *e.g.*, a carrier system. The SF tone is present in the idle state and absent during the seized state. In the seized state, dial pulses are conveyed by bursts of SF tone, corresponding to the interruptions in dc continuity created by a rotary dial or other dc dialing mechanism. *Note 2:* The SF tone may occupy a small portion of the user data channel spectrum, *e.g.,* 1600 Hz or 2600 Hz ("in-band" SF signaling), usually with a notch filter at the precise SF frequency, to prevent the user from inadvertently disconnecting a call if user data has a sufficiently strong spectral content at the SF frequency. The SF tone may also be just outside the user voice band, *e.g.,* 3600 Hz. *Note 3:* The Defense Data Network (DDN) transmits dc signaling pulses or supervisory signals, or both, over carrier channels or cable pairs on a 4-wire basis using a 2600-Hz signal tone. The conversion into tones, or vice versa, is done by SF signal units.

single-harmonic distortion. Of a fundamental frequency, the ratio of the power of a specified harmonic to the power of the fundamental frequency. *Note:* Single-harmonic distortion is measured at the output of a device under specified conditions and is expressed in dB.

single-mode fiber. *Synonym* single-mode optical fiber.

single-mode optical fiber. An optical fiber in which only the lowest order bound mode can propagate at the wavelength of interest. *Note 1:* The lowest order bound mode is ascertained for the wavelength of interest by solving Maxwell's equations for the boundary conditions imposed by the fiber, *e.g.*, core (spot) size and the refractive indices of the core and cladding. *Note 2:* The solution of Maxwell's equations for the lowest order bound mode will permit a pair of orthogonally polarized fields in the fiber, and this is the usual case in a communication fiber. *Note 3:* In step-index guides, single-mode operation occurs when the normalized frequency, *V*, is less than 2.405. For power-law profiles, single-mode operation occurs for a normalized frequency, *V*, less than approximately

$$2.405 \sqrt{\frac{g+2}{g}} ,$$

where *g* is the profile parameter. *Note 4:* In practice, the orthogonal polarizations may not be associated with degenerate modes. *Synonyms* monomode optical fiber, single-mode fiber, single-mode optical waveguide, unimode fiber.

single-mode optical waveguide. *Synonym* single-mode optical fiber.

single-Morse system. *Synonym* neutral direct-current telegraph system.

single-polarized antenna. An antenna that radiates or receives radio waves with a specific polarization. *Note:* For a singly polarized antenna, the desired sense of polarization is usually maintained only for certain directions or within the major portion of the radiation pattern.

single-sideband (SSB) emission. An amplitude modulated emission with one sideband only.

single-sideband (SSB) equipment reference level. The power of one of two equal tones that, when used together to modulate a transmitter, cause it to develop its full rated peak power output.

single-sideband suppressed carrier (SSB-SC) transmission. Single-sideband transmission in which the carrier is suppressed. *Note:* In SSB-SC the carrier power level is suppressed to the point where it is insufficient to demodulate the signal.

single-sideband (SSB) transmission. Sideband transmission in which only one sideband is transmitted. *Note:* The carrier may be suppressed.

single-tone interference. An undesired discrete frequency appearing in a transmission channel. *Note:* The single-tone interference frequency is the frequency that appears in the channel regardless of the nature of the source.

sink. 1. An absorber of energy. 2. In communications, a device that receives information, control, or other signals from a source.

S interface. For basic rate access in an Integrated Services Digital Network (ISDN) environment, a user-to-network interface reference point that (a) is characterized by a 4-wire, 144-kb/s (2B+D) user rate, (b) serves as a universal interface between ISDN terminals or terminal adapters and the network channel termination, (c) allows a variety of terminal types and subscriber networks, such as PBXs, local area networks (LANs), and controllers, to be connected to the network, and (d) operates at 4000 48-bit frames per second, *i.e.,* 192 kb/s, with a user portion of 36 bits per frame, *i.e.,* 144 kb/s.

six-bit byte. *Synonym* sextet.

skew. 1. In parallel transmission, the difference in arrival time of bits transmitted at the same time. 2. For data recorded on multichannel magnetic tape, the difference between reading times of bits recorded in a single transverse line. *Note:* Skew is usually interpreted to mean the difference in reading times between bits recorded on the tracks at the extremities, *i.e.,* edges, of the tape. 3. In facsimile systems, the angular deviation of the received frame from rectangularity caused by asynchronism between the scanner and the recorder. *Note:* Skew is expressed numerically as the tangent of the deviation angle. 4. In facsimile, the angle between the scanning line, or recording line, and the perpendicular to the paper path.

skew ray. In a multimode optical fiber, a bound ray that travels in a helical path along the fiber and thus (a) is not parallel to the fiber axis, (b) does not lie in a meridional plane, and (c) does not intersect the fiber axis.

skin effect. The tendency of alternating current to flow near the surface of a conductor, thereby restricting the current to a small part of the total cross-sectional area and increasing the resistance to the flow of current. *Note:* The skin effect is caused by the self-inductance of the conductor, which causes an increase in the inductive reactance at high frequencies, thus forcing the carriers, *i.e.,* electrons, toward the surface of the conductor. At high frequencies, the circumference is the preferred criterion for predicting resistance than is the cross-sectional area. The depth of penetration of current can be very small compared to the diameter.

skip distance. At a given azimuth, the minimum distance between the transmitting station and the closest point of return to the Earth of a transmitted wave reflected from the ionosphere.

skip zone. An annular region within the transmission range of an antenna, within which signals from the transmitter are not received. *Note:* The skip zone is bounded by the locus of the farthest points at which the ground wave can be received and the nearest points at which reflected sky waves can be received. *Synonyms* silent zone, zone of silence.

sky wave. A radio wave that travels upward from the antenna. *Note:* A sky wave may be reflected to Earth by the ionosphere.

slab-dielectric waveguide. An electromagnetic waveguide (a) that consists solely of dielectric materials, (b) in which the dielectric propagation medium has a rectangular cross section, (c) that has a width, thickness, and refractive indices that determine the operating wavelength and the modes the guide will support beyond the equilibrium length, (d) that may be cladded, protected, distributed, and electronically controllable, and (e) that may be used in various applications, such as in integrated optical circuits (IOCs) in which their shape is geometrically more convenient than the optical fibers that are circular in cross section, that are used in fiber optic cables for long-distance transmission. *Note:* Their principle of operation is the same as that for optical fibers that are circular in cross section. *Synonym* planar waveguide.

slant range. The line-of-sight distance between two points, not at the same level relative to a specific datum. *Note:* An example of slant range is the distance to an airborne radar target, *e.g.,* an airplane flying at high altitude with respect to that of the radar antenna. The slant range is the hypotenuse of the triangle represented by the altitude of the airplane and the distance between the radar antenna and the airplane's ground track (the point on the Earth at which it is directly overhead). In the absence of altitude information, the aircraft location would be plotted farther from the antenna than its actual ground track.

slave clock. A clock that is coordinated with a master clock. *Note 1:* Slave clock coordination is usually achieved by phase-locking the slave clock signal to a signal received from the master clock. *Note 2:* To adjust for the transit time of the signal from the master clock to the slave clock, the phase of the slave clock may be adjusted with respect to the signal from the master clock so that both clocks are in phase. Thus, the time markers of both clocks, at the output of the clocks, occur simultaneously.

slave station. 1. In a data network, a station that is selected and controlled by a master station. *Note:* Usually a slave station can only call, or be called by, a master station. 2. In navigation systems using precise time dissemination, a station having a clock is synchronized by a remote master station. *Synonym* subordinate station.

slewing. 1. Rotating a directional antenna or transducer rapidly about one or more axes. 2. Changing the frequency or pulse repetition rate of a signal source. 3. Changing the tuning of a receiver, usually by sweeping through many or all frequencies. 4. Redirecting the beam of a fixed antenna array by changing the relative phases of the signals feeding the antenna elements.

sliding window. A variable-duration window that allows a sender to transmit a specified number of data units before an acknowledgement is received or before a specified event occurs. *Note:* An example of a sliding window in packet transmission is one in which, after the sender fails to receive an acknowledgement for the first transmitted packet, the sender "slides" the window, *i.e.,* resets the window, and sends a second packet. This process is repeated for the specified number of times before the sender interrupts transmission. *Synonym (loosely)* acknowledgement delay period.

slip. In a sequence of transmitted symbols, *e.g.,* digital bits, a signal phase shift, *i.e.,* a signal positional displacement, that causes the loss of one or more symbols or the insertion of one or more extraneous symbols. *Note:* Slips are usually caused by inadequate synchronization of the two clocks controlling the transmission and reception of the signals that represent the symbols.

SLIP. *Acronym for* serial line Internet protocol. A protocol that allows a computer to use the Internet protocol (IP) with a standard telephone line and a high-speed modem.

slip-free operation. Operation of a communications system with sufficient phase-locking to avoid overflowing or emptying buffers.

slit source. *Synonym* line source.

slope. In a transmission line, the rate of change of attenuation with respect to frequency over the frequency spectrum. *Note 1:* The slope is usually expressed in dB per hertz or dB per octave. *Note 2:* In metallic lines, the slope is usually greater at high frequencies than at low frequencies.

slope equalizer. A device or circuit used to achieve a specified slope in a metallic transmission line.

slope-keypoint compaction. Data compaction accomplished by stating (a) a specific keypoint of departure, (b) a direction or slope of departure, (c) the maximum deviation from a prescribed specific value, and (d) a new keypoint and a new slope. *Note:* An example of slope-keypoint compaction is the storage or transmission of a slope and one point on a straight line instead of storing and transmitting a large number of values, *i.e.,* of points, on the line.

slot. In a distributed-queue dual-bus (DQDB) network, a protocol data unit (PDU) that (a) consists of 53 octets used to transfer segments of user information, (b) has the capacity to contain a segment of 52 octets and a 1-octet access control field, and (c) may be either a pre-arbitrated (PA) slot or a queued arbitrated (QA) slot.

slot antenna. A radiating element formed by a slot in a conducting surface or in the wall of a waveguide.

slotted-ring network. A ring network that allows unidirectional data transmission between data stations by transferring data in predefined slots in the transmission stream over one transmission medium such that the data return to the originating station.

slot time. In networks using carrier sense multiple access with collision detection (CSMA/CD), the length of time that a transmitting station waits before attempting to retransmit following a collision. *Note:* Slot time varies from station to station.

SMDR. *Abbreviation for* station message-detail recording.

smearing. In video displays, a localized distortion over a sub-region of the image, characterized by reduced sharpness of edges and spatial detail.

smooth Earth. Idealized surfaces, such as water surfaces or very level terrain, having radio horizons that are not formed by prominent ridges or mountains but are determined solely as a function of antenna height above ground and the effective Earth radius.

SNA. *Abbreviation for* systems network architecture. A proprietary communications architecture.

sneak current. In a communications circuit, an anomalous current that presents no immediate danger, but may cause improper operation or damage.

Snell's law. A law of geometric optics that defines the amount of bending that takes place when a light ray strikes a refractive boundary, *e.g.*, an air-glass interface, at a non-normal angle. *Note 1:* Snell's law states that

$$n_1 \sin\theta_1 = n_2 \sin\theta_2 ,$$

where n_1 is the index of refraction of the medium in which the incident ray travels, θ_1 is the angle, with respect to the normal at the refractive boundary, at which the incident ray strikes the boundary, n_2 is the index of refraction of the medium in which the refracted ray travels, and θ_2 is the angle, with respect to the normal at the refractive boundary, at which the refracted ray travels. The incident ray and refracted ray travel in the same plane, on opposite sides of the normal at the point of incidence. *Note 2:* If a ray travels from a medium of lower refractive index into a medium of higher refractive index, it is bent toward the normal; if it travels from a medium of higher refractive index to a medium of lower index, it is bent away from the normal. *Note 3:* If the incident ray travels in a medium of higher refractive index toward a medium of lower refractive index at such an angle that Snell's law would call for the sine of the refracted ray to be greater than unity (a mathematical impossibility); *i.e.*,

$$\sin\theta_2 = \frac{n_1}{n_2} \sin\theta_1 > 1,$$

then the "refracted" ray in actuality becomes a reflected ray and is totally reflected back into the medium of higher refractive index, at an angle equal to the incident angle (and thus still "obeys" Snell's Law). This reflection occurs even in the absence of a metallic reflective coating (*e.g.*, aluminum or silver). This phenomenon is called *"total internal reflection."* The smallest angle of incidence, with respect to the normal at the refractive boundary, which angle will support total internal reflection, is called the *"critical angle."*

SNMP. *Abbreviation for* Simple Network Management Protocol.

snow. In video display systems, noise that (a) is uniformly distributed on the display surface, such as that of a television or radar screen, (b) has the appearance of a uniform distribution of fixed or moving spots, mottling, or speckling, and (c) is usually caused by random noise on an intensity-modulated signal in a display device, such as a cathode-ray tube.

SNR. *Abbreviation for* signal-to-noise ratio.

soft copy. A nonpermanent display image, for example, a cathode ray tube display.

soft limiting. *See* limiting.

soft sectoring. On magnetic disks, magnetic drums, and optical disks, the identification of sector boundaries by using recorded information.

software. 1. A set of computer programs, procedures, and associated documentation concerned with the operation of a data processing system; *e.g.*, compilers, library routines, manuals, and circuit diagrams. 2. Information (generally copyrightable) that may provide instructions for computers; data for documentation; and voice, video, and music for entertainment or education.

software engineering. The discipline devoted to the design, development, and use of computer software. *Note:* Software engineering must address various aspects of data processing, including compatibility with the computer system which is to execute the software, and tradeoffs among maintainability, flexibility, efficiency, processing time, and costs.

software package. A package that consists of (a) one or more computer programs and possibly related material such as utility programs or tutorial programs, recorded on a medium suitable for delivery to the user, and from which the user can transfer the program(s) to a data-processing device, and (b) instructional materials such as handbooks and manuals, update information, and possibly support services information. *Note 1:* The computer programs may consist, for example, of application programs or operating systems, and are usually written in a high-level or low-level language, respectively. *Note 2:* The recording medium is usually a magnetic diskette or an optical compact disk.

software tool. Software, such as a computer program, routine, subroutine, program block, or program module, that can be used to develop, test, analyze, or maintain a computer program or its documentation. *Note:* Examples of software tools are automated software verification routines, compilers, program maintenance routines, bootstraps, program analyzers, and software monitors.

SOH. *Abbreviation for* start-of-heading character.

solid-state scanning. In facsimile, scanning in which all or a part of the scanning process is performed by electronic commutation of an array of solid-state photosensitive elements.

soliton. An optical pulse having a shape, spectral content, and power level designed to take advantage of nonlinear effects in an optical fiber waveguide, for the purpose of essentially negating dispersion over long distances.

sonar. *Acronym for* sound navigation and ranging. A device that is used primarily for the detection and location of underwater objects by reflecting acoustic waves from them, or by the interception of acoustic waves from an underwater, surface, or above-surface acoustic source. *Note:* Sonar operates with acoustic waves in the same way that radar and radio direction-finding equipment operate with electromagnetic waves, including use of the Doppler effect, radial component of velocity measurement, and triangulation.

SONET. *Acronym for* synchronous optical network. An interface standard for synchronous 2.46-Gb/s optical-fiber transmission, applicable to the Physical Layer of the OSI Reference Model. *Note 1:* SONET uses a basic data rate of 51.840 Mb/s, called OC1 (optical carrier 1). The SONET hierarchy is defined in multiples of OC1, up to and including OC48, for a maximum data rate of 2.48832 Gb/s. *Note 2:* SONET was developed by the Exchange Carriers Standards Association (ECSA).

sonobuoy. In sonar systems, a device (a) that is used to detect acoustic waves, such as those produced by ships and submarines, (b) that, when activated, relays information by radio, (c) that may be active or passive, and (d) that may be directional or nondirectional.

sounder prediction station. A station equipped with an ionosphere sounder for realtime monitoring of upper atmosphere phenomena or to obtain data for the prediction of propagation conditions.

sounding. In automated HF radio systems, the broadcasting of a very brief signal, containing the station address, station identifier, or call sign, to permit receiving stations to measure link quality.

sound navigation and ranging. *See* sonar.

sound-powered telephone. A telephone in which the operating power is derived from the speech input only.

sound wave. *See* acoustic wave.

source. In communications, that part of a system from which messages are considered to originate.

source efficiency. In optical systems, the ratio of emitted optical power of a source to the input electrical power.

source language. In computing, data processing, and communications systems, a language from which statements are translated. *Note:* Translators, assemblers, and compilers prepare target language programs, usually machine-language programs, from source language programs, usually high-level language programs written by programmers.

source program. 1. A computer program written in a source language. *Note:* An example of a source program is a program that serves as the input to an assembler, compiler, or translator. 2. A computer program that must be assembled, compiled, or translated before it can be executed by a computer.

source quench. A congestion-control technique in which a computer experiencing data traffic congestion sends a message back to the source of the messages or packets causing the congestion, requesting that the source stop transmitting.

source user. The user providing the information to be transferred to a destination user during a particular information transfer transaction. *Synonym* information source.

space. In telegraphy, one of the two significant conditions of encoding. *Note 1:* The complementary significant condition is called a *"mark." Note 2:* In modern digital communications, the two corresponding significant conditions of encoding are called *"zero"* and *"one."* *Synonyms* spacing pulse, spacing signal.

spacecraft. A man-made vehicle which is intended to go beyond the major portion of the Earth's atmosphere.

space diversity. A method of transmission or reception, or both, in which the effects of fading are minimized by the simultaneous use of two or more physically separated antennas, ideally separated by one or more wavelengths.

space-division multiplexing. *A misnomer. Note: Space-division multiplexing* has been improperly applied to the use of multiple physical transmission channels, *e.g.*, twisted pairs or optical fibers, under one sheath.

space-division switching. In telephony, switching in which single transmission-path routing determination is accomplished in a switch by using a physically separated set of matrix contacts or cross-points.

space operation service. A radiocommunication service concerned exclusively with the operation of spacecraft, in particular space tracking, space telemetry and space telecommand. These functions will normally be provided within the service in which the space station is operating.

space radiocommunication. Any radiocommunication involving the use of one or more space stations or the use of one or more reflecting satellites or other objects in space.

space research service. A radiocommunication service in which spacecraft or other objects in space are used for scientific or technological research purposes.

space station. A station located on an object which is beyond, is intended to go beyond, or has been beyond, the major portion of the Earth's atmosphere.

space subsystem. In satellite communications, that portion of the satellite link that is in orbit.

space system. Any group of cooperating Earth stations and/or space stations employing space radiocommunication for specific purposes.

space telecommand. The use of radiocommunication for the transmission of signals to a space station to initiate, modify or terminate functions of equipment on an associated space object, including the space station.

space telemetry. The use of telemetry for the transmission from a space station of results of measurements made in a spacecraft, including those relating to the functioning of spacecraft.

space tracking. Determination of the orbit, velocity or instantaneous position of an object in space by means of radiodetermination, excluding primary radar, for the purpose of following the movement of the object.

spacing bias. The uniform lengthening of all spacing signal pulses at the expense of the pulse width of all marking signal pulses.

spacing end distortion. *See* end distortion.

spacing pulse. *Synonym* space.

spacing signal. *Synonym* space.

spare. An individual part, subassembly, or assembly supplied for the maintenance or repair of systems or equipment.

spatial application. An application requiring high spatial resolution, possibly at the expense of reduced temporal positioning accuracy, *i.e.,* increased jerkiness. *Note:* Examples of spatial applications include the requirement to display small characters and to resolve fine detail in still video, or in motion video that contains very limited motion.

spatial coherence. *See* coherent.

spatial edge noise. In a video display, that form of edge busyness that is characterized by spatially varying distortion that occurs in close proximity to the edges of objects.

spatially coherent radiation. *See* coherent.

special grade access line. In the Defense Switched Network, an access line specially conditioned, usually by providing amplitude and delay equalization, to give it characteristics suitable for handling special services, such as reducing data signaling rates (DSR) to a rate between 600 b/s and 2400 b/s.

special grade of service. In the Defense Switched Network, a network-provided service in which specially conditioned interswitch trunks and access lines are used to provide secure voice, data, and facsimile transmission.

special interest group. *Synonym* community of interest.

specialized common carrier (SCC). A common carrier offering a limited type of service or serving a limited market.

special purpose computer. A computer that is designed to operate on a restricted class of problems.

special service. A radiocommunication service, not otherwise defined in this Section [of the *Radio Regulations*], carried on exclusively for specific needs of general utility, and not open to public correspondence. [RR with editor's note in brackets]

specification. 1. An essential technical requirement for items, materials, or services, including the procedures to be used to determine whether the requirement has been met. *Note:* Specifications may also include requirements for preservation, packaging, packing, and marking. 2. An official document intended primarily for supporting procurement, which document clearly and accurately describes the essential technical requirements for items, materials, or services, including the procedures by which it will be determined that the requirements have been met. *Note:* An example of a Federal specification is FIPS-PUB 159, *Detail Specification for 62.5-µm Core Diameter/125-µm Cladding Diameter Class Ia Multimode Optical Fibers.*

specific detectivity. For a photodetector, a figure of merit used to characterize performance, equal to the reciprocal of noise equivalent power (*NEP*), normalized to unit area and unit bandwidth. *Note:* Specific detectivity, D^*, is given by

$$D^* = \frac{\sqrt{A\Delta f}}{NEP},$$

where A is the area of the photosensitive region of the detector and Δf is the effective noise bandwidth. *Synonym* D-Star.

speckle noise. *Synonym* modal noise.

speckle pattern. In optical systems, a field-intensity pattern produced by the mutual interference of partially coherent beams that are subject to minute temporal and spatial fluctuations. *Note:* In a multimode fiber, a speckle pattern results from a superposition of mode field patterns. If the relative modal group velocities change with time, the speckle pattern will also change with time. If differential mode attenuation occurs, modal noise results.

spectral bandwidth. *See* spectral width.

spectral density. For a specified bandwidth of radiation consisting of a continuous frequency spectrum, the total power in the specified bandwidth divided by the specified bandwidth. *Note:* Spectral density is usually expressed in watts per hertz.

spectral irradiance. Irradiance per unit wavelength interval at a given wavelength, usually expressed in watts per unit area per unit wavelength interval.

spectral line. A narrow range of emitted or absorbed wavelengths.

spectral loss curve. Of an optical fiber, a plot of attenuation as a function of wavelength. *Note:* Spectral loss curves must be normalized with respect to distance before meaningful comparison among fibers can be made.

spectral purity. The degree to which a signal is monochromatic.

spectral radiance. Radiance per unit wavelength interval at a given wavelength, expressed in watts per steradian per unit area per wavelength interval.

spectral responsivity. The ratio of an optical detector's electrical output to its optical input, as a function of optical wavelength.

spectral width. The wavelength interval over which the magnitude of all spectral components is equal to or greater than a specified fraction of the magnitude of the component having the maximum value. *Note 1:* In optical communications applications, the usual method of specifying spectral width is the full width at half maximum. This method may be difficult to apply when the spectrum has a complex shape. Another method of specifying spectral width is a special case of root-mean-square deviation where the independent variable is wavelength, λ, and f (λ) is a suitable radiometric quantity. *Note 2:* The *relative spectral width*, Δλ/λ, is frequently used where Δλ is obtained according to note 1, and λ is the center wavelength.

spectral window. *See* window.

spectrum. *See* electromagnetic spectrum, optical spectrum.

spectrum designation of frequency. *See* electromagnetic spectrum.

spectrum signature. The pattern of radio signal frequencies, amplitudes, and phases, which pattern characterizes the output of a particular device and tends to distinguish it from other devices.

specular reflection. Reflection from a smooth surface, such as a mirror, which maintains the integrity of the incident wavefront.

speech digit signaling. *Synonym* bit robbing.

speech-plus. Pertaining to a circuit that was designed and used for speech transmission, but to which other uses, such as digital data transmission, facsimile transmission, telegraph, or signaling superimposed on the speech signals, have been added by means of multiplexing.

speech-plus-duplex operation. Operation in which speech and telegraphy (duplex or simplex) are transmitted simultaneously over the same circuit, and mutual interference is eliminated by the use of filters.

speech-plus-signaling. Pertaining to equipment that permits the use of part of a voice-frequency band for signaling.

speech power. *See* volume unit.

speech synthesizer. A device that is capable of accepting digital or analog data and developing intelligible speech sounds that correspond to the input data, without resorting to recorded sounds or without simply being a speech scrambler operating in reverse.

speed calling. A service feature that enables a switch or station to store certain telephone numbers and dial them automatically when a short (1-, 2-, or 3-digit) code is entered. *Contrast with* repertory dialer, speed dialing.

speed dialing. 1. *Synonym* abbreviated dialing. 2. Dialing at a speed greater than the normal ten pulses per second.

speed of light (*c*). The speed of an electromagnetic wave in free space, precisely 299,792,458 m/s. *Note 1:* The preceding figure is precise because by international agreement the meter is now defined in terms of the speed of light. *Note 2:* The speed of an electromagnetic wave, *e.g.*, light, is equal to the product of the wavelength and the frequency. *Note 3:* In any physical medium, the speed of light is lower than in free space. Since the frequency is not changed, the wavelength is also decreased.

speed of service. 1. The time between release of a message by the originator to receipt of the message by the addressee, as perceived by the end user. *Synonym* originator-to-recipient speed of service. 2. The time between entry of a message into a communications system and receipt of the message at the terminating communications facility, *i.e.,* the communications facility serving the addressee, as measured by the system.

speed-up tone. *Synonym* camp-on busy signal.

spike. An extremely short pulse of relatively high amplitude.

spike file. *See* last-in first-out.

spill forward. In automatic switching, the transfer of full control on a call to the succeeding office by sending forward the complete telephone address of the called party.

spill-forward feature. A service feature, in the operation of an intermediate office, that, acting on incoming trunk service treatment indications, assumes routing control of the call from the originating office. *Note:* This increases the chances of completion by offering the call to more trunk groups than are available in the originating office.

spillover. In an antenna, the part of the radiated energy from the feed that does not impinge on the reflectors.

spiral-four cable. A quadded cable with four conductors. *Synonym* star quadded cable.

splice. 1. To join, permanently, physical media that conduct or transmit power or a communication signal.

2. A device that so joins conducting or transmitting media. 3. The completed joint.

splice closure. A usually weatherproof encasement, commonly made of tough plastic, that envelops the exposed area between spliced cables, *i.e.,* where the jackets have been removed to expose the individual transmission media, optical or metallic, to be joined. *Note 1:* The closure usually contains some device or means to maintain continuity of the tensile strength members of the cables involved, and also may maintain electrical continuity of metallic armor, and/or provide external connectivity to such armor for electrical grounding. *Note 2:* In the case of fiber optic cables, it also contains a splice organizer to facilitate the splicing process and protect the exposed fibers from mechanical damage. *Note 3:* In addition to the seals at its seams and points of cable entry, the splice closure may be filled with an encapsulant to further retard the entry of water. *Synonym* closure.

splice loss. In fiber optic systems, any loss of optical power at a splice. *Note:* A practical splice, of physically realizable fibers, has losses attributable to a number of mechanisms, some of which are intrinsic to the fibers, and some of which are intrinsic to the method or device being used to join them.

splice organizer. In optical communication, a device that facilitates the splicing or breaking out of fiber optic cables. *Note:* The organizer provides means to separate and secure individual buffer tubes, fibers, and/or pigtails. It also provides means to secure mechanical splices or protective sleeves used in connection with fusion splices, and has means to contain the slack fiber that remains after the splicing process is completed.

split homing. The connection of a terminal facility to more than one switching center by separate access lines, each of which has a separate directory number.

split screen. On a display device, display space that has been divided into two or more areas, so that each area can display different portions of the same file or portions of different files. *Note 1:* The split screen excludes the data lying between the portions of the file or files being displayed and includes the desired data in the two or more windows afforded by the split screen. *Note 2:* Examples of split screens are screens in which different portions of a spreadsheet, database, graph, or picture that are too far apart in storage to be viewed or displayed simultaneously as a single image, are viewed adjacently on a single screen.

splitter. *See* directional coupler.

***(S+N)/N*.** *Abbreviation for* signal-plus-noise-to-noise ratio.

spontaneous emission. Radiation emitted when the internal energy of a quantum mechanical system drops from an excited level to a lower level without regard to the simultaneous presence of similar radiation. *Note:* Examples of spontaneous emission include radiation from an LED, and radiation from an injection laser below the lasing threshold.

spoofing. 1. (COMSEC) [The] interception, alteration, and retransmission of a cipher signal or data in such a way as to mislead the recipient. 2. (AIS) [An] attempt to gain access to an AIS by posing as an authorized user.

spooling. The use of auxiliary storage as buffer storage to reduce processing delays when transferring data between peripheral equipment and the processors of a computer. *Note:* The term is derived from the expression *"simultaneous peripheral operation on line."*

sporadic E. Irregular scattered patches of relatively dense ionization that develop seasonally within the E region and that reflect and scatter frequencies up to 150 MHz. *Note 1:* The sporadic E is a regular daytime occurrence over the equatorial regions and is common in the temperate latitudes in late spring, early summer and, to a lesser degree, in early winter. *Note 2:* At high, *i.e.,* polar, latitudes, sporadic E can accompany auroras and associated disturbed magnetic conditions. *Note 3:* The sporadic E can sometimes support reflections for distances up to 2,400 km at frequencies up to 150 MHz. *Synonym* sporadic E propagation.

sporadic E propagation. *Synonym* sporadic E.

spot beam. In satellite communications systems, a narrow beam from a satellite station antenna that illuminates, with high irradiance, a limited area of the Earth by using beam (directive) antennas rather than Earth-coverage antennas.

spot jamming. The jamming of a specific channel or frequency.

spot projection. In facsimile systems, optical scanning in which a scanning spot is moved across the object and the scanning spot size is determined by the illuminated area of the spot.

spot size. 1. The size of the electron spot on the face of a cathode ray tube. *Note:* The spot size is larger than the diameter of the electron beam because of the spillover of electrons into adjacent areas of the screen near the spot. The spot size is a function of the ability of the tube to focus the electron beam, as well as of the electron gun aperture. 2. In facsimile systems, the diameter of the scanning spot or the recording spot. 3. In single-mode optical fibers, the effective core diameter.

spot speed. In facsimile systems, the speed of the scanning or recording spot along the available line. *Note:* The spot speed is usually measured on the object or on the recorded copy.

spread spectrum. 1. Telecommunications techniques in which a signal is transmitted in a bandwidth considerably greater than the frequency content of the original information. *Note:* Frequency hopping, direct sequence spreading, time scrambling, and combinations of these techniques are forms of spread spectrum. 2. A signal structuring technique that employs direct sequence, frequency hopping or a hybrid of these, which can be used for multiple access and/or multiple functions. This technique decreases the potential interference to other receivers while achieving privacy and increasing the immunity of spread spectrum receivers to noise and interference. Spread spectrum generally makes use of a sequential noise-like signal structure to spread the normally narrowband information signal over a relatively wide band of frequencies. The receiver correlates the signals to retrieve the original information signal.

spur. A secondary route having a junction to the primary route in a network.

spurious emission. Emission on a frequency or frequencies which are outside the necessary bandwidth and the level of which may be reduced without affecting the corresponding transmission of information. Spurious emissions include harmonic emissions, parasitic emissions, intermodulation products and frequency conversion products, but exclude out-of-band emissions.

spurious radiation. Any unintentional emission.

spurious response. In radio reception, a response in the receiver intermediate frequency (IF) stage produced by an undesired emission in which the fundamental frequency (or harmonics above the fundamental frequency) of the undesired emission mixes with the fundamental or harmonic of the receiver local oscillator.

square wave. A wave that has two significant conditions, *i.e.,* two levels of amplitude, that change from one condition to the other in a relatively short time compared to the wavelength. *Note:* When the instantaneous amplitude is plotted versus time or distance, the waveform has a rectangular shape.

squelch. A circuit function that acts to suppress the audio output of a receiver. *Note:* The squelch function is activated in the absence of a sufficiently strong desired input signal, in order to exclude undesired lower-power input signals that may be present at or near the frequency of the desired signal. *Contrast with* noise suppression.

sr. *Abbreviation for* steradian.

SSB. *Abbreviation for* single sideband. *See* single-sideband emission.

SSB-SC. *Abbreviation for* single-sideband suppressed carrier. *See* single-sideband suppressed carrier transmission.

SS7. *Abbreviation for* Signaling System No. 7.

stability. The invariability of a specified property of a substance, device, or apparatus with time, or under the influence of typically extrinsic factors.

stagger. In facsimile systems, periodic error in the position of the recorded spot along the recorded line.

standard. 1. Guideline documentation that reflects agreements on products, practices, or operations by nationally or internationally recognized industrial, professional, trade associations or governmental bodies. *Note:* This concept applies to formal, approved standards, as contrasted to de facto standards and proprietary standards, which are exceptions to this concept. 2. An exact value, a physical entity, or an abstract concept, established and defined by authority, custom, or common consent to serve as a reference, model, or rule in measuring quantities or qualities, establishing practices or procedures, or evaluating results. A fixed quantity or quality.

standard frequency and time signal-satellite service. A radiocommunication service using space stations on Earth satellites for the same purpose as those of the standard frequency and time signal service. This service may also include feeder links necessary for its operation.

standard frequency and time signal service. A radiocommunication service for scientific, technical and other purposes, providing the transmission of specified frequencies, time signals, or both, of stated high precision, intended for general reception.

standard frequency and time signal station. A station in the standard frequency and time signal service.

Standard Generalized Mark-up Language. *See* SGML.

standardized profile. A profile that specifies one or more interoperable open systems interconnection stacks that are intended to cover one or more specific functional areas. *Note:* Examples of standardized profiles are the ISO standardized profiles and the NATO standardized profiles.

standard optical source. A reference optical source to which emitting and detecting devices are compared for calibration purposes. *Note:* In the United States, recognized standard optical sources must be traceable to the National Institute of Standards and Technology (NIST), formerly the National Bureau of Standards (NBS).

standard telegraph level (*STL*). The power per individual telegraph channel required to yield the standard composite data level. *Note:* For example, for a composite data level of -13 dBm at 0-dBm transmission level point (0TLP), the *STL* would be approximately -25 dBm for a 16-channel VFCT terminal computed from $STL = -(13+10\log_{10} n)$, where *n* is the number of telegraph channels and the *STL* is in dBm.

standard test signal. A single-frequency signal with standardized level used for testing the peak power transmission capability and for measuring the total harmonic distortion of circuits or parts of a circuit. *Note:* Standardized test signal levels and frequencies are listed in MIL-STD-188-100 and in the *Code of Federal Regulations,* Title 47, part 68.

standard test tone. A single-frequency signal with a standardized level generally used for level alignment of single links and of links in tandem. *Note:* For standardized test signal levels and frequencies, see MIL-STD-188-100 for DOD use, and the *Code of Federal Regulations*, Title 47, part 68 for other Government agencies.

standard time and frequency signal (STFS) service. In the United States, standard time and frequency signals, broadcast on very precise carrier frequencies by the U.S. Naval Observatory and the National Institute of Standards and Technology (NIST), formerly the National Bureau of Standards (NBS). *Note:* The *Radio Regulations* (RR) define an identical international service as standard frequency and time signal service.

standby. 1. In computer and communications systems operations, pertaining to a power-saving condition or status of operation of equipment that is ready for use but not in use. *Note:* An example of a standby condition is a radio station operating condition in which the operator can receive but is not transmitting. 2. Pertaining to a dormant operating condition or state of a system or equipment that permits complete resumption of operation in a stable state within a short time. 3. Pertaining to spare equipment that is placed in operation only when other, in-use equipment becomes inoperative. *Note:* Standby equipment is usually classified as (a) *hot* standby equipment, which is warmed up, *i.e.,* powered and ready for immediate service, and which may be switched into service automatically upon detection of a failure in the regular equipment, or (b) *cold* standby equipment, which is turned off or not connected to a primary power source, and which must be placed into service manually.

standing wave. In a transmission line, a wave in which the distribution of current, voltage, or field strength is formed by the superposition of two waves propagating in opposite directions, and which wave is characterized by a series of nodes (maxima) and anti-nodes (minima) at fixed points along the transmission line. *Note:* A standing wave may be formed when a wave is transmitted into one end of a transmission line and is reflected from the other end by an impedance mismatch, *i.e.*, discontinuity, such as an open or a short. *Synonym* stationary wave.

standing wave ratio (*SWR*). The ratio of the amplitude of a standing wave at an anti-node (minimum) to the amplitude at an adjacent node (maximum). *Note 1:* The standing wave ratio (*SWR*) in a uniform transmission line is given by

$$SWR = \frac{1+\rho}{1-\rho},$$

where ρ is the reflection coefficient. *Note 2:* Reflections occur as a result of discontinuities, such as an imperfection in an otherwise uniform transmission line, or when a transmission line is terminated with other than its characteristic impedance.

star coupler. A passive optical coupler having a number of input and output ports, used in network applications. *Note:* An optical signal introduced into any input port is distributed to all output ports. Because of the nature of the construction of a passive star coupler, the number of ports is usually a power of 2; *i.e.*, two input ports and two output ports (a "two-port" coupler, customarily called a *"directional coupler,"* or *"splitter"*); four input ports and four output ports (a "four-port" coupler); eight input ports and eight output ports (an "eight-port" coupler); *etc.*

star network. *See* network topology.

star quadded cable. *Synonym* spiral-four cable.

starting frame delimiter. A specified bit pattern that indicates the start of a transmission frame.

start message. *Synonym* go-ahead notice.

start notice. *Synonym* go-ahead notice.

start-of-heading character (SOH). A transmission control character used as the first character of a message heading.

start-of-text character (STX). A transmission control character that precedes a text and may be used to terminate the message heading.

star topology. *See* network topology.

start pulse. *See* A-condition, start signal.

start-record signal. In facsimile systems, a signal used for starting the process of converting the electrical signal to an image on the record medium.

start signal. 1. A signal that prepares a device to receive data or to perform a function. *Contrast with* **A-condition.** 2. In start-stop transmission, a signal at the beginning of a character that prepares the receiving device for the reception of the code elements. *Note:* A start signal is limited to one signal element usually having the duration of a unit interval.

start-stop character. A character that includes one start signal at the beginning and one or two stop signals at the end.

start-stop distortion. In start-stop modulation, the ratio of (a) the maximum absolute difference between the actual and the theoretical intervals that separate any significant instant of modulation or demodulation from the significant instant of the start signal element immediately preceding it to (b) the unit interval.

start-stop margin. In start-stop modulation, the maximum amount of overall start-stop distortion that is compatible with correct translation by the start-stop equipment of all the character signals that appear singly, that appear at the maximum allowable speed, or that appear at the standard modulation rate.

start-stop modulation. A method of modulation in which the time of occurrence of the bits within each character, or block of characters, relates to a fixed time frame, but the start of each character, or block of characters, is not related to this fixed time frame.

start-stop system. *Synonym* asynchronous communications system.

start-stop transmission. 1. Asynchronous transmission in which a start pulse and a stop pulse are used for each symbol. 2. Signaling in which each group of code elements corresponding to an alphanumeric character is (a) preceded by a start signal that serves to prepare the receiving mechanism for the reception and registration of a character and (b) followed by a stop signal that serves to bring the receiving mechanism to rest in preparation for the reception of the next character.

start-stop TTY distortion. *Synonym* teletypewriter signal distortion.

statement. 1. In programming languages, a language construct that represents a set of declarations or a step in a sequence of actions. 2. In computer programming, a symbol string or other arrangement of symbols. 3. In computer programming, a meaningful expression or generalized instruction, represented in a source language.

staticizer. *See* serial-to-parallel conversion.

station. One or more transmitters or receivers or a combination of transmitters and receivers, including the accessory equipment, necessary at one location for carrying on a radiocommunication service, or the radio astronomy service. Each station shall be classified by the service in which it operates permanently or temporarily.

stationary satellite. *See* geostationary orbit.

stationary wave. *Synonym* standing wave.

station battery. Within a facility, a separate battery power source that satisfies all significant requirements for dc input power associated with the facility. *Note:* Station batteries are usually centrally located. The batteries may power radio and telephone equipment as well as provide emergency lighting and controls for the equipment.

station clock. In a station, the principal clock, or alternate clock, that provides the timing reference at the station.

station equipment. *See* customer premises equipment.

station load. The total power requirements of the integrated station facilities.

station message-detail recording (SMDR). A record of all calls originated or received by a switching system. *Note:* SMDRs are usually generated by a computer.

statistical multiplexing. Multiplexing in which channels are established on a statistical basis; *i.e.*, connections are made according to probability of need.

statistical time-division multiplexing. Time-division multiplexing in which connections to communication circuits are made on a statistical basis.

statute mile. A unit of distance equal to 1.609 km (0.869 nmi, 5280 ft.).

STDM. *Abbreviation for* statistical time-division multiplexing.

steady-state condition. 1. In a communications circuit, a condition in which some specified characteristic of a condition, such as a value, rate, periodicity, or amplitude, exhibits only negligible change over an arbitrarily long period. 2. In an electrical circuit, the condition that exists after all initial transients or fluctuating conditions have damped out, and all currents, voltages, or fields remain essentially constant, or oscillate uniformly. 3. In fiber optics, *synonym for* equilibrium mode distribution.

step-by-step (SXS) switching system. An automatic dial telephone system in which calls are switched by a succession of switches that move a step at a time, from stage to stage, each step being made in response to the dialing of a number.

step-index fiber. An optical fiber with a core having a uniform refractive index.

step-index profile. For an optical fiber, a refractive index profile characterized by a uniform refractive index within the core and a sharp decrease in refractive index at the core-cladding interface. *Note 1:* The step-index profile corresponds to a power-law index profile with the profile parameter approaching infinity. *Note 2:* The step-index profile is used in most single-mode fibers and some multimode fibers.

steradian (sr). The metric unit of solid angle. *See* metric system.

stereophonic crosstalk. An undesired signal occurring in the main channel from modulation of the stereophonic channel or that occurring in the stereophonic channel from modulation of the main channel.

stereophonic sound subcarrier. A subcarrier within the FM broadcast baseband used for transmitting signals for stereophonic sound reception of the main broadcast program service.

STFS. *Abbreviation for* standard time and frequency signal. *See* standard time and frequency signal service.

still image. Nonmoving visual information, *i.e.,* fixed images, such as graphs, drawings, and pictures.

still video. Video imagery that is not intended to convey the appearance of movement. *Contrast with* freeze frame, freeze frame television.

stimulated emission. In a quantum mechanical system, the radiation emitted when the internal energy of the system drops from an excited level (induced by the presence of radiant energy at the same frequency) to a lower level. *Note:* An example of stimulated emission is the radiation from an injection laser diode operated above the lasing threshold.

STL. *Abbreviation for* standard telegraph level, studio-to-transmitter link.

stopband. A band of frequencies, between specified limits, that a circuit, such as a filter or telephone circuit, does not transmit. *Note 1:* Frequencies above the lower limit and below the upper limit are not transmitted, *i.e.*, are not allowed to pass. *Note 2:* The limiting frequencies are those at which the transmitted power level increases to a specified level, usually 3 dB below the maximum level, as the frequency is decreased or increased from that at which the transmitted power is a minimum. *Note 3:* The difference between the limits is the stopband bandwidth, usually expressed in hertz.

stop element. *See* stop signal.

stop-record signal. In facsimile systems, a signal used for stopping the process of converting the electrical signal to an image on the record medium.

stop signal. 1. In start-stop transmission, a signal at the end of a character that prepares the receiving device for the reception of a subsequent character. A stop signal is usually limited to one signal element having any duration equal to or greater than a specified minimum value. 2. A signal to a receiving mechanism to wait for the next signal.

storage. 1. The retention of data in any form, usually for the purpose of orderly retrieval and documentation. 2. A device consisting of electronic, electrostatic, electrical, hardware or other elements into which data may be entered, and from which data may be obtained, as desired.

storage cell. 1. An addressable storage unit. 2. The smallest subdivision of storage into which a unit of data can be entered, stored, and retrieved. *Synonym* storage element.

storage element. *Synonym* storage cell.

storage register. *See* register.

store-and-forward (S-F). Pertaining to communications systems in which messages are received at intermediate routing points and recorded *i.e.,* stored, and then transmitted, *i.e.,* forwarded, to the next routing point or to the ultimate recipient.

store-and-forward switching center. A message switching center in which a message is accepted from the originating user, *i.e.,* sender, when it is offered, held in a physical storage, and forwarded to the destination user, *i.e.,* receiver, in accordance with the priority placed upon the message by the originating user and the availability of an outgoing channel.

stored-program computer. A computer that (a) is controlled by internally stored instructions, (b) can synthesize and store instructions, and (c) can subsequently execute those instructions.

STP. *Abbreviation for* signal transfer point.

strap. *See* cross-connection.

stray current. Electrical current through a path other than the intended path.

streamer. *Synonym* streaming tape drive.

streaming tape drive. A magnetic tape unit capable of recording from, and dumping to, another storage medium without stopping at interblock gaps. *Note:* Streaming tape drives are often used for bulk transfer of data between tape and disk storage. *Synonym* streamer.

streaming tape recording. A method of recording on magnetic tape, which method maintains continuous tape motion without the requirement to start and stop within the interrecord gap.

strength member. Any component of a communication cable, metallic or optical, the function of which is to protect the transport medium, *i.e.,* conductor or fiber, from excessive tensile and bending stresses during installation and while in service.

stressed environment. In radiocommunications, an environment that is under the influence of extrinsic factors that degrade communications integrity, such as when (a) the benign communications medium is disturbed by natural or man-made events (such as an intentional nuclear burst), (b) the received signal is degraded by natural or man-made interference (such as jamming signals or co-channel interference), (c) an interfering signal can reconfigure the network, and/or (d) an adversary threatens successful communications, in which case radio signals may be encrypted in order to deny the adversary an intelligible message, traffic flow information, network information, or automatic link establishment (ALE) control information.

string. A sequence of data elements, such as bits or characters, considered as a whole.

stroke. A straight line or arc that is used as a segment of a graphic character.

stroke edge. In character recognition, the line of discontinuity between a side of a stroke and the background, obtained by averaging, over the length of the stroke, the irregularities resulting from the printing and detecting processes.

stroke speed. In facsimile systems, the rate at which a fixed line perpendicular to the direction of scanning is crossed in one direction by a scanning or recording spot. *Note 1:* Stroke speed is usually expressed as a number of strokes per minute. When the system scans in both directions, the stroke speed is twice this number. *Note 2:* In most conventional mechanical systems, the stroke speed is equivalent to drum speed.

stroke width. In character recognition, the distance between the two edges of a stroke, measured perpendicular to the stroke centerline.

structured programming. A technique for organizing and coding computer programs in which a hierarchy of modules is used, each having a single entry and a single exit point, and in which control is passed downward through the structure without unconditional branches to higher levels of the structure. Three types of control flow are used: sequential, test, and iteration.

STU. *Acronym for* secure telephone unit. A U.S. Government-approved telecommunications terminal that protects the transmission of sensitive or classified information in voice, data, and facsimile systems.

studio-to-transmitter link (STL). A communications link used for the transmission of broadcast material from a studio to the transmitter. *Note:* The STL may be a microwave, radio, or landline link.

stuffing. *See* bit stuffing, de-stuffing.

stunt box. A device that controls the nonprinting functions of a printer at a terminal.

STX. *Abbreviation for* start-of-text character.

SUB. *Acronym for* **su**bstitute character.

sub-band adaptive differential pulse code modulation (SB-ADPCM). Modulation in which (a) an audio frequency band is split into two sub-bands, *i.e.*, a higher and a lower band, and (b) the signals in each sub-band are encoded using ADPCM.

subcarrier. A carrier used to modulate another carrier. *Note:* The modulated carrier can be used to modulate another carrier, and so on, so that there can be several levels of subcarriers, *i.e.*, several intermediate carriers.

sublayer. 1. In a layered open communications system, a specified subset of the services, functions, and protocols included in a given layer. 2. In the Open Systems Interconnection–Reference Model, a subdivision of a given layer, *e.g.*, a conceptually complete group of the services, functions, and protocols included in the given layer.

subnet address. In an Internet Protocol (IP) address, an extension that allows users in a network to use a single IP network address for multiple physical subnetworks. *Note:* The IP address contains three parts: the network, the subnet, and host addresses. Inside the subnetwork, gateways and hosts divide the local portion of the IP address into a subnet address and a host address. Outside of the subnetwork, routing continues as usual by dividing the destination address into a network portion and a local portion.

subnetwork. A collection of equipment and physical transmission media that forms an autonomous whole and that can be used to interconnect systems for purposes of communication.

subordinate station. *Synonym* slave station.

subroutine. A set of computer instructions to carry out a predefined function or computation. *Note: "Open"* subroutines are integrated into the main program.

"Closed" subroutines are arranged so that program control is shifted to them for execution of their task(s) and then returned to the main program.

subscriber. In a public switched telecommunications network, the ultimate user, *i.e.,* customer, of a communications service. *Note 1:* Subscribers include individuals, activities, organizations, etc. *Note 2:* Subscribers use end instruments, such as telephones, modems, facsimile machines, computers, and remote terminals, that are connected to a central office. *Note 3:* Subscribers are usually subject to tariff. *Note 4:* Subscribers do not include communications systems operating personnel except for their personal terminals.

subscriber line. *Synonym* loop (def. #1).

substitute character (SUB). A control character that is used in the place of a character that is recognized to be invalid or in error or that cannot be represented on a given device.

substitution method. In optical fiber technology, a method of measuring the transmission loss by (a) using a stable optical source, at the wavelength of interest, to drive a mode scrambler, the output of which overfills (drives) a 1-meter to 2-meter reference fiber having physical and optical characteristics matching those of the fiber under test, (b) measuring the power level at the output of the reference fiber, (c) repeating the procedure, substituting the fiber under test for the reference fiber, and (d) subtracting the power level obtained at the output of the fiber under test from the power level obtained at the output of the reference fiber, to get the transmission loss of the fiber under test. *Note 1:* The substitution method has certain shortcomings with regard to its accuracy, but its simplicity makes it a popular field test method. It is conservative, in that if it were used to measure the individual losses of several long fibers, and the long fibers were concatenated, the total loss obtained (excluding splice losses) would be expected to be lower than the sum of the individual fiber losses. *Note 2:* Some modern optical power meters have the capability to set to zero the reference level measured at the output of the reference fiber, so that the transmission loss of the fiber under test may be read out directly.

subvoice-grade channel. A channel with a bandwidth narrower than that of a voice-grade channel. *Note:* A subvoice-grade channel is usually a subchannel of a voice-grade line.

successful block delivery. The transfer of a nonduplicate user information block between the source user and intended destination user. *Note:* Successful block delivery includes the delivery of correct and incorrect blocks. *Contrast with* successful block transfer.

successful block transfer. The transfer of a correct, nonduplicate, user information block between the source user and intended destination user. *Note:* Successful block transfer occurs when the last bit of the transferred block crosses the functional interface between the telecommunications system and the intended destination user. Successful block transfer can only occur within a defined maximum block transfer time after initiation of a block transfer attempt. *Contrast with* successful block delivery.

successful disengagement. The termination of user information transfer between a source user and a destination user in response to a disengagement request. *Note:* Successful disengagement occurs at the earliest moment at which either user is able to initiate a new information transfer transaction.

sudden ionospheric disturbance (SID). An abnormally high ionization density in the D region caused by an occasional sudden solar flare, *i.e.*, outburst of ultraviolet light from the Sun. *Note:* The SID results in a sudden increase in radio-wave absorption that is most severe in the upper medium-frequency (MF) and lower high-frequency (HF) ranges.

sum check. *Synonym* summation check.

summation check. 1. A check based on the formation of the sum of the digits of a numeral. *Note:* The sum of the individual digits is usually compared with a previously computed value. 2. A comparison of checksums on the same data on different occasions or on different representations of the data in order to verify data integrity. *Synonym* sum check.

sunspot. In the photosphere, *i.e.*, visible disk of the Sun, a dark marking that manifests a magnetic anomaly that is associated with interference with radio communications on Earth. *Note:* Sunspot activity, *i.e.*, the number of sunspots occurring at a given time or on a given day, is cyclic. The period of a cycle, from maximum through minimum and back to maximum sunspot count, is approximately 11 years.

superencryption. [The] process of encrypting encrypted information. *Note:* [This process] occurs when a message, encrypted off-line, is transmitted over a secured, on-line circuit, or when information encrypted by the originator is multiplexed into a communications trunk, which is then bulk encrypted.

supergroup. *See* group, multiplex hierarchy.

supergroup distribution frame (SGDF). In frequency-division multiplexing (FDM), the distribution frame that provides terminating and interconnecting facilities for group modulator output, group demodulator input, supergroup modulator input, and supergroup demodulator output circuits of the basic supergroup spectrum of 312 kHz to 552 kHz.

super high frequency (SHF). *See* electromagnetic spectrum.

superluminescent LED. A light-emitting diode in which there is stimulated emission with amplification but insufficient feedback for oscillations to build up to achieve lasing action.

superradiance. In a gain medium, amplification of spontaneously emitted radiation characterized by moderate spectral line narrowing and moderate directionality. *Note:* Superradiance is usually distinguished from lasing action by the absence of positive feedback, and hence the absence of well-defined modes of oscillation.

supervisor. *Synonym* supervisory program.

supervisory control. The use of characters or signals for the automatic actuation of equipment or indicators.

supervisory program. 1. A program, usually part of an operating system, that controls the execution of other routines and regulates work scheduling, input-output operations, error actions, and similar functions. 2. A program that allocates computer component space and schedules computer events by task queuing and system interrupts. *Note:* Control of the system is returned to the supervisory program frequently enough to ensure that demands on the system are met. *Synonym* supervisory routine. 3. A computer program, usually part of an operating system, that controls the execution of other computer programs and regulates the flow of work in a data processing system. *Synonyms* executive program, supervisor.

supervisory routine. *Synonym* supervisory program.

supervisory signals. Signals used to indicate, or to indicate and control, the various operating states of the circuits or circuit combinations involved in a particular connection.

suppressed carrier single-sideband emission. A single-sideband emission in which the carrier is virtually suppressed and not intended to be used for demodulation.

suppressed carrier transmission. Amplitude modulation (AM) transmission in which the carrier level is reduced below that required for demodulation. *Note 1:* Reduction of the carrier level permits higher power levels in the sidebands than would be possible with conventional AM transmission. *Note 2:* Carrier power must be restored by the receiving station to permit demodulation. *Note 3:* Suppressed carrier transmission is a special case of reduced carrier transmission.

surface refractivity. The refractive index of the Earth's atmosphere, calculated from observations of pressure, temperature, and humidity at the surface of the Earth. *Note:* The surface refractivity gradient is the difference in refractive index between the surface and a given altitude, such as between the surface and 1000 m.

surface wave. A wave that is guided along the interface between two different media or by a refractive index gradient. *Note 1:* The field components of the wave diminish with distance from the interface. *Note 2:* Optical energy is not converted from the surface wave field to another form of energy and the wave does not have a component directed normal to the interface surface. *Note 3:* In optical fiber transmission, evanescent waves are surface waves. *Note 4:* In radio transmission, ground waves are surface waves that propagate close to the surface of the Earth, the Earth having one refractive index and the atmosphere another, thus constituting an interface surface.

surge. *Synonym* impulse.

surge suppressor. *Synonym* arrester.

survey. *See* path survey.

survivability. A property of a system, subsystem, equipment, process, or procedure that provides a defined degree of assurance that the named entity will continue to function during and after a natural or man-made disturbance; *e.g.,* nuclear burst. *Note:* For a given application, survivability must be qualified by specifying the range of conditions over which the entity will survive, the minimum acceptable level or post-disturbance functionality, and the maximum acceptable outage duration.

survivable operation. *See* survivability.

survival craft station. A mobile station in the maritime mobile service or the aeronautical mobile service intended solely for survival purposes and located on any lifeboat, life-raft or other survival equipment.

susceptibility. In electronic warfare, the degree to which electronic equipment is affected by electromagnetic energy radiated by an enemy's equipment, such as jamming transmitters.

susceptibility threshold. The amount of undesired signal power required at the input terminals of a receiver to cause barely perceptible interference at the receiver output terminals.

susceptiveness. In telephone systems, the extent to which circuits pick up noise and low-frequency energy by induction from power systems. *Note:* Susceptiveness depends on telephone circuit balance, wire and connection transpositions, wire spacing, and isolation from ground.

sweep acquisition. A technique whereby the frequency of the local oscillator is slowly swept past the reference in order to assure that the pull-in range is reached.

sweep jamming. Jamming in which (a) a narrow frequency band of jamming energy is repeatedly swept over a relatively wide frequency band, (b) the sweep rate is such as to be on any given frequency only long enough to accomplish its jamming task, returning to that frequency again before the expiration of the jammed circuit recovery time. *Note 1:* Sweep jamming combines the advantages of both spot- and barrage-jamming by rapid electronic sweeping of a narrow band of jamming signals over a broad frequency spectrum. *Note 2:* The disadvantage of sweep-jamming is its high susceptibility to electronic counter-countermeasures.

swim. Slow, graceful, undesired movements of display elements, groups, or images about their mean position on a display surface, such as that of a monitor. *Note 1:* Swim can be followed by the human eye, whereas jitter usually appears as a blur. *Note 2:* Jitter, swim, wander, and drift have increasing periods of variation in that order.

switch. 1. In communications systems, a mechanical, electro-mechanical, or electronic device for making, breaking, or changing the connections in or among circuits. 2. *Deprecated synonym for* central office, switching center. 3. In communications systems, to transfer a connection from one circuit to another. 4. In a computer program, a conditional instruction and a flag that is interrogated by the instruction. 5. In a computer program, a parameter that controls branching and that is bound, prior to the branch point being reached. *Synonym* switchpoint. 6. In computer programming, a programming technique or statement for making a selection, such as a conditional jump. 7. In computer software applications, a functional unit, such as a toggle button, used to make selections.

switchboard. Equipment used for manual switching operations.

switch busy hour. In telephony, the busy hour for a single switch.

switched circuit. In a communications network, a circuit that may be temporarily established at the request of one or more of the connected stations.

switched loop. In telephony, a circuit that automatically releases a connection from a console or switchboard, once the connection has been made to the appropriate terminal. *Note:* Loop buttons or jacks are used to answer incoming listed directory number calls, dial "0" internal calls, transfer requests, and intercepted calls. The attendant can handle only one call at a time. *Synonym* released loop.

switched multimegabit data services (SMDS). . A connectionless, broadband, packet-switched data service that provides LAN-like performance and features in metropolitan or wide areas. *Note:* Currently SMDS operates at 1.544 Mb/s (megabits per second) or 44.736 Mb/s. These are the T1 and T3 rates, respectively, over switched fiber optic networks.

switched network. 1. A communications network, such as the public switched telephone network, in which any user may be connected to any other user through the use of message, circuit, or packet switching and control devices. 2. Any network providing switched communications service.

switching. The controlling or routing of signals in circuits to execute logical or arithmetic operations or to transmit data between specific points in a network. *Note:* Switching may be performed by electronic, optical, or electromechanical devices.

switching center. In communications systems, a facility in which switches are used to interconnect communications circuits on a circuit-, message-, or packet-switching basis. *Synonyms, in telephony,* central office, switching exchange, switching facility. *Deprecated synonym* **switch.**

switching exchange. *Synonym* switching center.

switching facility. *Synonym* switching center.

switching system. 1. A communications system consisting of switching centers and their interconnecting media. 2. Part of a communication system organized to temporarily associate functional units, transmission channels or telecommunication circuits for the purpose of providing a desired telecommunication facility. *Note:* Examples of NATO-owned switching system are IVSN and TARE.

switchpoint. *Synonym* switch (def. #5).

SWR. *Abbreviation for* standing wave ratio.

SX. *Abbreviation for* simplex signaling.

SXS. *Abbreviation for* step-by-step switching system.

syllable. A character string or a bit string in a word.

symbolic language. A computer programming language used to express addresses and instructions with symbols convenient to humans rather than to machines.

symbolic logic. The discipline in which valid arguments and operations are dealt with using an artificial language designed to avoid the ambiguities and logical inadequacies of natural languages.

symmetrical channel. A channel in which the send and receive circuits have the same data signaling rate.

symmetrical pair. A balanced transmission line, in a multipair cable, having equal conductor resistances per unit length, equal impedances from each conductor to earth, and equal impedances to other lines.

SYN. *Acronym for* synchronous idle character.

sync pulse. *Synonym* synchronization pulse.

synchronism. 1. The state of being synchronous. 2. For repetitive events with the same, multiple, or submultiple repetition rates, a relationship among the events such that a significant instant of one event bears a fixed time relationship to a corresponding instant in another event. *Note:* Synchronism is maintained when there is a fixed, *i.e.*, constant, phase relationship among the group of repetitive events. 3. The simultaneous occurrence of two or more events at the same instant on the same coordinated time scale.

synchronization. 1. The attaining of synchronism. 2. The obtaining of a desired fixed relationship among corresponding significant instants of two or more signals. 3. A state of simultaneous occurrences of significant instants among two or more signals.

synchronization bit. A bit used to achieve or maintain synchronism. *Note:* The term *"synchronization bit"* is usually applied to digital data streams, whereas the term *"synchronization pulse"* is usually applied to analog signals.

synchronization code. In digital systems, a sequence of bits introduced into a transmitted signal to achieve or maintain synchronism.

synchronization pulse. A pulse used to achieve or maintain synchronism. *Note:* The term *"synchronization pulse"* is usually applied to analog signals, whereas the term *"synchronization bit"* is usually applied to digital data streams. *Synonym* sync pulse.

synchronizing. 1. Achieving and maintaining synchronism. 2. In facsimile, achieving and maintaining predetermined speed relations between the scanning spot and the recording spot within each scanning line. *Note:* In the civilian community, the noun *"synchronization"* is preferred to *"synchronizing."*

synchronizing pilot. In FDM, a reference frequency used for achieving and maintaining syntonization of the oscillators of a carrier system or for comparing the frequencies or phases of the signals generated by those oscillators.

synchronizing signal. In facsimile systems, the signal that maintains predetermined speed relations between the scanning spot and recording spot within each facsimile scanning line.

synchronous. 1. Pertaining to the relationship of two or more repetitive signals that have simultaneous occurrences of significant instants. *Note:* "*Isochronous*" and "*anisochronous*" pertain to characteristics. "*Synchronous*" and "*asynchronous*" pertain to relationships. 2. Pertaining to synchronism (def.#2).

synchronous crypto-operation. [A] method of on-line crypto-operation in which crypto-equipment and associated terminals have timing systems to keep them in step.

synchronous data link control (SDLC). In a data network, a bit-oriented protocol for the control of synchronous transmission over data links.

synchronous data network. A data network in which synchronism is achieved and maintained between data circuit-terminating equipment (DCE) and the data switching exchange (DSE), and between DSEs. *Note:* The data signaling rates are controlled by timing equipment within the network.

synchronous height. *See* synchronous orbit.

synchronous idle character (SYN). A transmission control character used in synchronous transmission systems to provide a signal from which synchronism or synchronous correction may be achieved between data terminal equipment, particularly when no other character is being transmitted.

synchronous network. A network in which clocks are controlled to run, ideally, at identical rates, or at the same mean rate with a fixed relative phase displacement, within a specified limited range. *Note:* Ideally, the clocks are synchronous, but they may be mesochronous in practice. By common usage, such mesochronous networks are frequently described as *"synchronous."*

synchronous optical network. *See* SONET.

synchronous orbit. Any orbit in which an orbiting object has a period equal to the average rotational period of the body being orbited, and in the same direction of rotation as that body. *Note 1:* A synchronous orbit need not be equatorial, but it usually is, ideally. A body in a nonequatorial synchronous orbit will, when observed from a fixed point on the orbited body, appear to move up and down, *i.e.,* northward and southward. If the synchronous orbit is not perfectly circular, the orbiting body will appear to move back and forth, eastward and westward. The combination of these two motions will produce a figure-8 pattern as seen from the orbited body. *Note 2:* A synchronous orbit about the Earth that is circular and lies in the equatorial plane is called a geostationary orbit.

synchronous satellite. A satellite in a synchronous orbit.

synchronous system. A system in which events, such as signals, occur in synchronism. *Note:* An example of a synchronous system is one in which a transmitter and receiver operate with a fixed time relationship.

synchronous TDM. A multiplexing scheme in which timing is obtained from a clock that controls both the multiplexer and the channel source.

synchronous transfer mode. In a Broadband Integrated Services Digital Network (B-ISDN), a proposed transport level technique in which time-division multiplexing and switching is to be used across the user's network interface.

synchronous transmission. Digital transmission in which the time interval between any two similar significant instants in the overall bit stream is always an integral number of unit intervals. *Note:* "*Isochronous*" and "*anisochronous*" are characteristics, while "*synchronous*" and "*asynchronous*" pertain to relationships.

syntax. 1. In a language, the relationships among characters or groups of characters, independent of their meanings or the manner of their interpretation and use. 2. The structure of expressions in a language. 3. The rules governing the structure of a language. 4. In a language, the relationship among symbols. *Note:* In computer languages, as in all artificial languages, syntax is developed, and usually described, before their use begins. In natural languages, syntax is developed, and sometimes never described, after use has begun.

syntonization. The process of setting the frequency of one oscillator equal to that of another.

SYSGEN. *Acronym for* system generation.

system. 1. Any organized assembly of resources and procedures united and regulated by interaction or interdependence to accomplish a set of specific functions. 2. A collection of personnel, equipment, and methods organized to accomplish a set of specific functions.

system administration. In computer technology, a set of functions that provides support services, ensures reliable operations, promotes efficient use of the system, and ensures that prescribed service-quality objectives are met. *Synonym* system management.

system analysis. A systematic investigation of a real or planned system to determine the functions of the system and how they relate to each other and to any other system. *Synonym* **systems analysis.**

system blocking. *Synonym* access denial.

system blocking signal. A control message generated within a telecommunications system to indicate temporary unavailability of system resources required to complete a requested access. *Note:* The system blocking signal is part of system overhead information.

system budget. *See* power budget.

system documentation. The collection of documents that describes the requirements, capabilities, limitations, design, operation, and maintenance of a system, such as a communications, computing, or information processing system.

system failure transfer. In the event of a catastrophic failure, the ability to transfer central office trunks or interoffice trunking to predetermined stations to allow incoming and outgoing calls to be completed.

system follow-up. The study of the effects of a system after it has reached a stabilized state of operational use. *Synonyms* post-development review, post-implementation review.

system generation (SYSGEN). The process of selecting optional parts of an operating system and of creating a particular operating system tailored to the requirements of a data processing installation.

system integration. The progressive linking and testing of system components to merge their functional and technical characteristics into a comprehensive, interoperable system. *Note:* Integration of data systems allows data existing on disparate systems to be shared or accessed across functional or system boundaries.

system integrity. 1. That condition of a system wherein its mandated operational and technical parameters are within the prescribed limits. 2. [The] quality of an AIS when it performs its intended function in an unimpaired manner, free from deliberate or inadvertent unauthorized manipulation of the system.

system lifecycle. The course of developmental changes through which a system passes from its conception to the termination of its use and subsequent salvage. *Note:* For example, a system lifecycle might include the phases and activities associated with the analysis, acquisition, design, development, test, integration, operation, maintenance, and modification of the system.

system loading. In a frequency-division multiplexed (FDM) transmission system, the absolute power level of the composite signal transmitted in one direction. *Note 1:* The absolute power level is referred to a zero transmission level point (0TLP). *Note 2:* The composite signal contains signaling, speech, and digital signals.

system management. 1. Network management functions extended to include subscriber elements or user end instruments. 2. In computer systems, *synonym* system administration.

system operational threshold. For a supported performance parameter of a system, the value that establishes the minimum operational service performance level for the parameter. *Note:* A measured parameter value worse than the system operational threshold indicates that the system is in an outage state.

system overhead information. *See* overhead information.

system power margin. *Synonym* power margin.

system reliability. The probability that a system, including all hardware, firmware, and software, will satisfactorily perform the task for which it was designed or intended, for a specified time and in a specified environment.

system robustness. The measure or extent of the ability of a system, such as a computer, communications, data processing, or weapons system, to continue to function despite the existence of faults in its component subsystems or parts. *Note:* System performance may be diminished or otherwise altered until the faults are corrected.

systems analysis. *Synonym* system analysis.

systems control. In a communications system, the control and implementation of a set of functions that (a) prevent or eliminate degradation of any part of the system, (b) initiate immediate response to demands that are placed on the system, (c) respond to changes in the system to meet long range requirements, and (d) may include various subfunctions, such as (i) immediate circuit utilization actions, (ii) continuous control of circuit quality, (iii) continuous control of equipment performance, (iv) development of procedures for immediate repair, restoration, or replacement of facilities and equipment, (v) continuous liaison with system users and with representatives of other systems, and (vi) the provision of advice and assistance in system use.

systems design. 1. A process of defining the hardware and software architecture, components, modules, interfaces, and data for a system to satisfy specified requirements. 2. The preparation of an assembly of methods, procedures, or techniques united by regulated interaction to form an organized whole.

systems engineering. *See* systems design.

system signaling and supervision. In transmission systems, any scheme used to provide such functions as system control, addressing, routing, error detection and correction, level control, priority, traffic control, message accountability, and/or other required overhead information.

system software. Application-independent software that supports the running of application software.

system standard. In the military community, the system-specific characteristics, not dictated by the individual components' electrical performance characteristics, but necessary in order to permit internal and external interoperability.

system supervision. In telephone systems, the use of signals and techniques to perform system management functions, such as system control, addressing, routing, error detection and correction, level control, priority, traffic control, message accountability, and other overhead functions that may be described in system overhead portions of messages.

system support. The continued provision of services and material necessary for the use and improvement of a system during its lifecycle.

system test time. The part of operating time during which a functional unit is tested for proper operation. *Note:* In a computer, the system test time may include the time for testing programs belonging to the operating system.

T. *Abbreviation for* tera (10^{12}). *See* International System of Units.

Tactical Automatic Digital Switching System (TADSS). A transportable store-and-forward message-switching system used for rapid deployment in support of tactical forces.

tactical command and control (C^2) systems. The equipment, communications, procedures, and personnel essential to a commander for planning, directing, coordinating, and controlling tactical operations of assigned forces pursuant to assigned missions.

tactical communications. Communications in which information of any kind, especially orders and decisions, are conveyed from one command, person, or place to another within the tactical forces, usually by means of electronic equipment, including communications security equipment, organic to the tactical forces. *Note:* Tactical communications do not include communications provided to tactical forces by the Defense Communications System (DCS), to nontactical military commands, and to tactical forces by civil organizations.

tactical communications system. A communications system that (a) is used within, or in direct support of, tactical forces, (b) is designed to meet the requirements of changing tactical situations and varying environmental conditions, (c) provides securable communications, such as voice, data, and video, among mobile users to facilitate command and control within, and in support of, tactical forces, and (d) usually requires extremely short installation times, usually on the order of hours, in order to meet the requirements of frequent relocation.

tactical data information link (TADIL). A standardized communications link, approved by the Joint Staff, that is suitable for transmission of digital information, and is characterized by standardized message formats and transmission characteristics.

tactical data information link–A (TADIL–A). A netted link in which one unit acts as a net control station and interrogates each unit by roll call. *Note:* Once interrogated, that unit transmits its data to the net. This means that each unit receives all the information transmitted. This is a direct transfer of data and no relaying is involved.

tactical data information link–B (TADIL–B). A point-to-point data link between two units which provides for simultaneous transmission and reception of data (duplex).

tactical load. For the host service tactical forces, the total power requirements for communications, including the requirements for weapons, detection, command and control systems, and related support functions. *Note:* The tactical load is a part of the operational load.

TADIL. *Acronym for* tactical data information link.

TADSS. *Acronym for* Tactical Automatic Digital Switching System.

tag. *See* flag, label.

tag image file format (TIFF). A file format used to store an image using the particular data structure of the file.

TAI. *Abbreviation for* International Atomic Time.

tail circuit. A communications line from the end of a major transmission link, such as a microwave link, satellite link, or LAN, to the end-user location. *Note:* A tail circuit is a part of a user-to-user connection.

tailing. In facsimile systems, the excessive prolongation of the decay of the signal. *Synonym* **hangover**.

takeoff angle. *Synonym* departure angle.

tandem. Pertaining to an arrangement or sequencing of networks, circuits, or links, in which the output terminals of one network, circuit, or link are connected directly to the input terminals of another network, circuit, or link. *Note:* For example, concatenated microwave links constitute a tandem connection.

tandem center. In a switched public telecommunications network, a facility that connects trunks to trunks and does not connect any local loops.

tandem office. A central office that serves local subscriber loops, and also is used as an intermediate switching point for traffic between central offices.

tandem tie trunk network (TTTN). An arrangement that permits sequential connection of tie trunks between PBX and Centrex® locations by using tandem operation. *Note:* Tandem operation permits two or more dial tie trunks to be connected at a tandem center to form a through connection.

tap. 1. To draw energy from a circuit. 2. To monitor, with or without authorization, the information that is being transmitted via a communications circuit. 3. To extract a portion of the signal from an optical fiber or communications link. *Note:* One method of tapping an optical fiber is to bend it to a relatively short radius, thus promoting radiation of a portion of the optical signal.

tapered fiber. An optical fiber in which the cross section, *i.e.,* cross-sectional diameter or area, varies, *i.e.,* increases or decreases, monotonically with length.

tape relay. A method of retransmitting TTY traffic from one channel to another, in which messages arriving on an incoming channel are recorded in the form of perforated tape, this tape then being either fed directly and automatically into an outgoing channel, or manually transferred to an automatic transmitter for transmission on an outgoing channel.

target language. In computing, data processing, and communications systems, a language into which statements are translated. *Note:* Translators, assemblers, and compilers prepare target language programs, usually machine-language programs, from source language programs, usually high-level language programs written by programmers.

tariff. The published schedule of rates or charges for a specific unit of equipment, facility, or type of service such as might be provided by a telecommunications common carrier.

TASI. *Acronym for* time-assignment speech interpolation.

tasking. *See* multitasking.

TAT. *Abbreviation for* trans-Atlantic telecommunications (cable). *Note: TAT* formerly stood for *transatlantic telephone (cable).*

T-carrier. The generic designator for any of several digitally multiplexed telecommunications carrier systems. *Note 1:* The designators for T-carrier in the North American digital hierarchy correspond to the designators for the digital signal (DS) level hierarchy. *See the table on the following page. Note 2:* T-carrier systems were originally designed to transmit digitized voice signals. Current applications also include digital data transmission. *Note 3:* If an "F" precedes the "T", a fiber optic cable system is indicated at the same rates. *Note 4:* The table below lists the designators and rates for current T-Carrier systems. *Note 5:* The North American and Japanese hierarchies are based on multiplexing 24 voice-frequency channels and multiples thereof, whereas the European hierarchy is based on multiplexing 30 voice-frequency channels and multiples thereof. *See table on following page.*

TCB. *Abbreviation for* trusted computing base.

TCF. *Abbreviation for* technical control facility.

T-coupler. A passive optical coupler having three ports (three fibers). *Note 1:* Two isolated inputs may be combined into one output; or one input, into two isolated outputs. *Note 2:* The amount of coupling loss, usually expressed in dB, between ports is determined by the design and construction of the coupler. *Synonym* splitter.

TCP. *Abbreviation for* Transmission Control Protocol. In the Internet Protocol suite, a standard, connection-oriented, full-duplex, host-to-host protocol used over packet-switched computer communications networks. *Note 1:* TCP corresponds closely to the ISO Open Systems Interconnection—Reference Model (OSI—RM) Layer 4 (Transport Layer). *Note 2:* The OSI—RM uses TP-0 or TP-4 protocols for transmission control.

TCP/IP. *Abbreviation for* Transmission Control Protocol/ Internet Protocol. Two interrelated protocols that are part of the Internet protocol suite. *Note 1:* TCP operates on the OSI Transport Layer and breaks data into packets. IP operates on the OSI Network Layer and routes packets. *Note 2:* TCP/IP was originally developed by the U.S. Department of Defense.

TCP/IP Suite. The suite of interrelated protocols associated with Transmission Control Protocol/Internet Protocol. *Note 1:* The TCP/IP Suite includes, but is not limited to, protocols such as TCP, IP, UDP, ICMP, FTP, and SMTP. *Note 2:* Additional application and management protocols are sometimes considered part of the TCP/IP Suite. This includes protocols such as SNMP.

TCS. *Abbreviation for* trusted computer system.

TCU. *Abbreviation for* teletypewriter control unit.

TDD. *Abbreviation for* Telecommunications Device for the Deaf.

TDM. *Abbreviation for* time division multiplex.

TDMA. *Abbreviation for* time-division multiple access.

TE. *Abbreviation for* transverse electric. *See* transverse electric mode.

technical area. In the military community, an area in which temperature, humidity, or access is controlled because it contains equipment, such as communications, computing, control, or support equipment, that requires such controls.

technical control facility (TCF). A physical plant, or a designated and specially configured part thereof, that (a) contains the equipment necessary for ensuring fast, reliable, and secure exchange of information, (b) typically includes distribution frames and associated panels, jacks, and switches and monitoring, test, conditioning, and orderwire equipment, and (c) allows telecommunications systems control personnel to exercise operational control of communications paths and facilities, make quality analyses of communications and communications channels, monitor operations and maintenance functions, recognize and correct deteriorating conditions, restore disrupted communications, provide requested on-call circuits, and take or direct such actions as may be required and practical to provide effective telecommunications services.

technical control hubbing repeater. *Synonym* data conferencing repeater.

technical load. The portion of the operational load required for communications, tactical operations, and ancillary equipment including necessary lighting, air-conditioning, or ventilation required for full continuity of communications.

technical vulnerability. In information handling, a hardware, software, or firmware weakness, or design deficiency, that leaves a system open to assault, harm, or unauthorized exploitation, either externally or internally, thereby resulting in unacceptable risk of information compromise, information alteration, or service denial.

TED. *Abbreviation for* trunk encryption device.

tee coupler. A passive coupler that has three ports.

TEK. *Abbreviation for* traffic encryption key.

teleaction service. In Integrated Services Digital Network (ISDN) applications, a telecommunications service that uses very short messages with very low data transmission rates between the user and the network.

telecommand. The use of telecommunication for the transmission of signals to initiate, modify or terminate functions of equipment at a distance.

telecommunication. 1. Any transmission, emission, or reception of signs, signals, writing, images and sounds or intelligence of any nature by wire, radio, optical or other electromagnetic systems. 2. Any transmission, emission, or reception of signs, signals, writings, images, sounds, or information of any nature by wire, radio, visual, or other electromagnetic systems.

telecommunication architecture. *See* network architecture.

telecommunications center. *See* communications center.

Telecommunications Device for the Deaf (TDD). A machine that uses typed input and output, usually with a visual text display, to enable individuals with hearing or speech impairments to communicate over a telecommunications network.

telecommunications facilities. The aggregate of equipment, such as radios, telephones, teletypewriters, facsimile equipment, data equipment, cables, and switches, used for providing telecommunications services.

telecommunications management network (TMN). A network that interfaces with a telecommunications network at several points in order to receive information from, and to control the operation of, the telecommunications network. *Note:* A TMN may use parts of the managed telecommunications network to provide for the TMN communications.

telecommunications security. *See* communications security.

telecommunications service. 1. Any service provided by a telecommunication provider. 2. A specified set of user-information transfer capabilities provided to a group of users by a telecommunications system. *Note:* The telecommunications service user is responsible for the information content of the message. The telecommunications service provider has the responsibility for the acceptance, transmission, and delivery of the message.

Telecommunications Service Priority (TSP) service. A regulated service provided by a telecommunications provider, such as an operating telephone company or a carrier, for NS/EP telecommunications. *Note:* The TSP service replaced Restoration Priority (RP) service effective September 1990.

Telecommunications Service Priority (TSP) system. A system that provides a means for telecommunications users to obtain priority treatment from service providers for the NS/EP telecommunications requirements. *Note:* The TSP system replaced the Restoration Priority (RP) system effective September 1990.

Telecommunications Service Priority (TSP) system user. Any individual, organization, or activity that interacts with the NS/EP TSP System.

telecommunications system. *See* **communications** system.

telecommunications system operator. The organization responsible for providing telecommunications services to users.

teleconference. The live exchange of information among persons and machines remote from one another but linked by a telecommunications system. *Note:* The telecommunications system may support the teleconference by providing audio, video, and data services by one or more means, such as telephone, telegraph, teletype, radio, and television.

telegram. Written matter intended to be transmitted by telegraphy for delivery to the addressee. This term also includes radiotelegrams unless otherwise specified. In this definition the term telegraphy has the same general meaning as defined in the [1979 General Worldwide Administrative Radio Conference] Convention.

telegraph. *See* telegraphy.

telegraphy. A form of telecommunication which is concerned in any process providing transmission and reproduction at a distance of documentary matter, such as written or printed matter or fixed images, or the reproduction at a distance of any kind of information in such a form. For the purposes of the *Radio Regulations*, unless otherwise specified therein, telegraphy shall mean a form of telecommunication for the transmission of written matter by the use of a signal code.

telemetry. The use of telecommunication for automatically indicating or recording measurements at a distance from the measuring instrument.

telephone. A user end instrument that is used to transmit and receive voice-frequency signals.

telephone exchange. *Synonym* **central** office.

telephone frequency. *See* audio frequency, voice frequency.

telephone number. The unique network address that is assigned to a telephone user, *i.e.*, subscriber, for routing telephone calls.

telephone sidetone. *Synonym* sidetone.

telephony. 1. The branch of science devoted to the transmission, reception, and reproduction of sounds, such as speech and tones that represent digits for signaling. *Note 1:* Transmission may be via various media, such as wire, optical fibers, or radio. *Note 2:*

Analog representations of sounds may be digitized, transmitted, and, on reception, converted back to analog form. *Note 3: "Telephony"* originally entailed only the transmission of voice and voice-frequency data. Currently, it includes new services, such as the transmission of graphics information. 2. A form of telecommunication set up for the transmission of speech or, in some cases, other sounds.

telephoto. Pertaining to pictures transmitted via a telecommunications system.

teleprinter. A teletypewriter that can only receive data and does not have a keyboard for transmission.

teleprocessing. The combining of telecommunications and computer operations interacting in the automatic processing, reception, and transmission of data and/or information. *Note:* Teleprocessing includes human-machine interface equipment.

teleseminar. *See* teletraining.

teleservice. *See* telecommunications service.

teletex. An international store-and-forward essentially error-free communications service that is defined by the CCITT, has a data signaling rate (DSR) of 2400 b/s over switched telephone networks, and has a communications protocol that supports the CCITT Group 4 facsimile service.

teletext. A type of one-way information service in which a subscriber can receive data on a video display. *Note:* The information is transmitted to the subscriber's video display over a common carrier channel. A proprietary video adapter unit is required for reception. *Contrast with* viewdata.

teletraining. Training that (a) in which usually live instruction is conveyed in real time via telecommunications facilities, (b) that may be accomplished on a point-to-point basis or on a point-to-multipoint basis, and (c) may assume many forms, such as a teleseminar, a teleconference, or an electronic classroom, usually including both audio and video. *Synonyms* distance learning, distance training, electronic classroom, virtual instruction.

teletypewriter (TTY). A printing telegraph instrument that has a signal-actuated mechanism for automatically printing received messages. *Note 1:* A TTY may have a keyboard similar to that of a typewriter for sending messages. *Note 2:* Radio circuits carrying TTY traffic are called "RTTY circuits" or "RATT circuits."

teletypewriter control unit (TCU). A device that controls and coordinates operations between teletypewriters and message switching centers.

teletypewriter exchange service (TWX). A switched teletypewriter service in which suitably arranged teletypewriter stations are provided with lines to a central office for access to other such stations.

teletypewriter signal distortion. The shifting of signal pulse transitions from their proper positions relative to the beginning of the start pulse. *Note:* The magnitude of the distortion is expressed in percent of a perfect unit pulse length. *Synonym* start-stop TTY distortion.

television (TV). A form of telecommunication for the transmission of transient images of fixed or moving objects. *Note 1:* The picture signal is usually accompanied by the sound signal. *Note 2:* In North America, TV signals are generated, transmitted, received, and displayed in accordance with the NTSC standard.

television broadcast translator. *See* translator (def. #3).

Telex®. A communication service involving teletypewriters connected through automatic exchanges.

Telnet. The TCP/IP standard network virtual terminal protocol that is used for remote terminal connection service and that allows a user at one site to interact with systems at other sites as if that user terminal were directly connected to computers at those sites.

TEM. *Abbreviation for* transverse electric and magnetic.

TEMPEST. 1. [A] Short name referring to investigation, study, and control of compromising emanations from telecommunications and automated information systems equipment. 2. To shield against compromising emanations.

temporal application. A video application requiring high temporal resolution, *i.e.,* reduced jerkiness, possibly at the expense of reduced spatial resolution. *Note:* An example of temporal applications is the ability to accurately discern moving image features such as facial expressions and lip movements.

temporal coherence. *See* coherent.

temporal edge noise. In a video display, that form of edge busyness that is characterized by time-varying sharpness at the edges of objects.

temporally coherent radiation. *See* coherence time.

terahertz (THz). A unit denoting one trillion (10^{12}) hertz.

terminal. A device capable of sending, receiving, or sending and receiving information over a communications channel.

Terminal Access Controller (TAC). A host computer that accepts terminal connections, usually from dial-up lines, and that allows the user to invoke Internet remote log-on procedures, such as Telnet.

terminal adapter. An interfacing device employed at the "R" reference point in an ISDN environment that allows connection of a non-ISDN terminal at the physical layer to communicate with an ISDN network. *Note:* Typically, a terminal adapter will support standard RJ-11 telephone connection plugs for voice and RS-232C, V.35 and RS-449 interfaces for data.

terminal endpoint (TE) functional group. A functional group that includes functions broadly belonging to Layer 1 and higher layers of the CCITT Recommendation X.200 Reference Model. *Note 1:* The functions of a TE functional group are performed on various types of equipment, or combinations of equipment, such as digital telephones, data terminal equipment, and/or integrated work stations. *Note 2:* Examples of TE functions are protocol-handling, maintenance, interface, and connection functions.

terminal equipment. 1. Communications equipment at either end of a communications link, used to permit the stations involved to accomplish the mission for which the link was established. 2. In radio-relay systems, equipment used at points where data are inserted or derived, as distinct from equipment used only to relay a reconstituted signal. 3. Telephone and telegraph switchboards and other centrally located equipment at which communications circuits are terminated.

terminal impedance. 1. The impedance as measured at the unloaded output terminals of transmission equipment or a line that is otherwise in normal operating condition. 2. The ratio of voltage to current at the output terminals of a device, including the connected load.

terminal mobility. In commercial wireless networks, the ability of a terminal, while in motion, to access telecommunication services from different locations, and the capability of the network to identify and locate that terminal.

terminal mobility management. In personal communications service (PCS), (a) providing authentication of terminal information, (b) maintaining terminal location and capability information for each terminal, and (c) providing translation between terminal identification and location (routing address) for the completion of calls to terminals.

termination. 1. The load connected to a transmission line, circuit, or device. *Note:* For a uniform transmission line, if the termination impedance is equal to the characteristic impedance of the line, wave reflections from the end of the line will be avoided. 2. In hollow metallic waveguides, the point at which energy propagating in the waveguide continues in a nonwaveguide propagation mode into a load. 3. An impedance, often resistive, that is connected to a transmission line or piece of equipment as a dummy load, for test purposes.

terminus. A device used to terminate, position, and hold an optical fiber within a connector.

ternary signal. A signal that can assume, at any given instant, one of three significant conditions, such as power level, phase position, pulse duration, or frequency. *Note:* Examples of ternary signals are (a) a pulse that can have a positive, zero, or negative voltage value at any given instant, (b) a sine wave that can assume phases of 0°, 120°, or 240° relative to a clock pulse, and (c) a carrier wave that can assume any one of three different frequencies depending on three different modulation signal significant conditions.

terrestrial radiocommunication. Any radiocommunication other than space radiocommunication or radio astronomy.

terrestrial station. A station effecting terrestrial radiocommunication. In these *[Radio] Regulations,* unless otherwise stated, any station is a terrestrial station.

test and validation. Physical measurements taken (a) to verify conclusions obtained from mathematical modeling and analysis or (b) for the purpose of developing mathematical models.

test antenna. An antenna of known performance characteristics used in determining transmission characteristics of equipment and associated propagation paths.

test center. *See* patch and test facility, technical control facility.

test point. A point within a piece of equipment or an equipment string that provides access to signals for the purpose of fault isolation.

test tone. A tone sent at a predetermined level and frequency through a transmission system for test purposes, such as for facilitating measurements and for aligning gains and losses in the system.

text processing. *Synonym* word processing.

T4 (carrier). *See* T-carrier.

T5 (carrier). *See* T-carrier.

TG. *Abbreviation for* telegraph. *See* telegraphy.

TGM. *Abbreviation for* trunk group multiplexer.

THD. *Abbreviation for* total harmonic distortion.

thermal noise. The noise generated by thermal agitation of electrons in a conductor. The noise power, *P*, in watts, is given by *P* = k*T*Ä*f*, where k is Boltzmann's constant in joules per kelvin, *T* is the conductor temperature in kelvins, and Ä*f* is the bandwidth in hertz. *Note 1:* Thermal noise power, per hertz, is equal throughout the frequency spectrum, depending only on k and *T*. *Note 2:* For the general case, the above definition may be held to apply to charge carriers in any type of conducting medium. *Synonym* Johnson noise.

thermal radiation. 1. Electromagnetic radiations emitted from a heat or light source as a consequence of its temperature; it consists essentially of ultraviolet, visible, and infrared radiations. 2. The heat and light produced by a nuclear explosion.

thermodynamic temperature. A measure, in kelvins (K), proportional to the thermal energy of a given body at equilibrium. *Note 1:* A temperature of 0 K is called "absolute zero," and coincides with the minimum molecular activity (*i.e.*, thermal energy) of matter. *Note 2:* Thermodynamic temperature was formerly called "absolute temperature." *Note 3:* In practice, the International Temperature Scale of 1990 (ITS-90) serves as the basis for high-accuracy temperature measurements in science and technology.

THF. *Abbreviation for* **tre**mendously high frequency. *See* electromagnetic spectrum.

thin-film laser. A laser that is constructed by thin-film deposition techniques on a substrate for use as a light source, is usually used to drive thin-film optical waveguides, and may be used in integrated optical circuits.

thin-film optical modulator. A modulator that consists of multilayered films of material of different optical characteristics, is capable of modulating transmitted light by using electro-optic, electro-acoustic, or magneto-optic effects to obtain signal modulation, and may be used as a component in integrated optical circuits.

thin-film optical multiplexer. A multiplexer that consists of multilayered films of material of different optical characteristics, is capable of multiplexing transmitted light by using electro-optic, electro-acoustic, or magneto-optic effects to obtain signal multiplexing, and may be used as a component in integrated optical circuits.

thin-film optical switch. A switch that consists of multilayered films of material of different optical characteristics, that is capable of switching transmitted light by using electro-optic, electro-acoustic, or magneto-optic effects to obtain signal switching, and is usually used as a component in integrated optical circuits. *Note:* Thin-film optical switches may support only one propagation mode.

thin-film optical waveguide. A slab-dielectric waveguide that consists of multilayered films of material of different optical characteristics, is capable of guiding an optical signal, and may be used as a component in integrated optical circuits.

third-order intercept point. A point (a) that is an extrapolated convergence—not directly measurable—of intermodulation distortion products in the desired output and (b) that indicates how well a receiver performs in the presence of strong nearby signals. *Note:* Determination of a third-order intercept point is accomplished by using two test frequencies that fall within the first intermediate frequency mixer passband. Usually, the test frequencies are about 20 to 30 kHz apart.

third window. Of silica-based optical fibers, the transmission window at approximately 1.55 μm. *Note:* The third window is the minimum-loss window in silica-based fibers.

threat. Capabilities, intentions, and attack methods of adversaries to exploit, or any circumstance or event with the potential to cause harm to, information or an information system.

three-bit byte. *Synonym* triplet.

three-way calling. A switching system service feature that permits users to add a third party at a different number during a call, without the assistance of an attendant.

threshold. 1. The minimum value of a signal that can be detected by the system or sensor under consideration. 2. A value used to denote predetermined levels, such as those pertaining to volume of message storage, *i.e.,* in-transit storage or queue storage, used in a message switching center. 3. The minimum value of the parameter used to activate a device. 4. The minimum value a stimulus may have to create a desired effect.

threshold current. In a laser, the driving current corresponding to lasing threshold.

threshold extension. *See* FM threshold extension.

threshold frequency. In optoelectronics, the frequency of incident radiant energy below which there is no photoemissive effect.

through group. A group of 12 voice-frequency channels transmitted as a unit through a carrier system.

through-group equipment. In carrier telephone transmission, equipment that accepts the signal from a group receiver output and attenuates it to the proper signal level for insertion, without frequency translation, at the input of a group transmitter.

throughput. 1. The number of bits, characters, or blocks passing through a data communication system, or portion of that system. *Note 1:* Throughput may vary greatly from its theoretical maximum. *Note 2:* Throughput is expressed in data units per period of time; *e.g.,* in the DDN, as blocks per second. 2. The maximum capacity of a communications channel or system. 3. A measure of the amount of work performed by a system over a period of time, *e.g.,* the number of jobs per day.

through supergroup. An aggregate of 60 voice-frequency channels, *i.e.,* five groups, transmitted as a unit through a carrier system.

through-supergroup equipment. In carrier telephone transmission, equipment that accepts the multiplexed signal from a supergroup receiver output, amplifies it without frequency translation, and provides the proper signal level to the input of a supergroup transmitter equipment.

THz. *Abbreviation for* terahertz. *See* International System of Units.

TIA. *Abbreviation for* Telecommunications Industry Association. *See* EIA interface.

ticketed call. A call for which a record is made of certain facts concerning the call, such as the time it was placed, the duration, the call originator, call destination numbers, and, where applicable, the attendant's name or initials.

TIE. *Acronym for* time interval error.

tie line. *See* tie trunk.

tie trunk. A telephone line that directly connects two private branch exchanges (PBXs).

TIFF. *Acronym for* tagged image file format.

tight buffer. *See* buffer.

tiling. *See* block distortion.

time. 1. An epoch, *i.e.,* the designation of an instant on a selected time scale, astronomical or atomic. It is used in the sense of time of day 2. On a time scale, the interval between two events, or the duration of an event. 3. An apparently irreversible continuum of ordered events.

time ambiguity. A situation in which more than one different time or time measurement can be obtained under the stated conditions.

time-assignment speech interpolation (TASI). An analog technique used on certain long transmission links to increase voice-transmission capacity. *Note:* TASI works by switching additional users onto any channel temporarily idled because an original user has stopped speaking. When the original user resumes speaking, that user will, in turn, be switched to any channel that happens to be idle.

time availability. *Synonym* circuit reliability.

time block. An arbitrary grouping of several consecutive hours of a day, usually for a particular season, during which it is assumed that propagation data are statistically homogeneous.

time code. A code used for the transmission and identification of time signals. *Note:* In telecommunications systems, the format of the time code must be specified.

time code ambiguity. The shortest interval between successive repetitions of the same time code value. *Note:* For example, in a time code in which year-of-century is the most slowly changing field, the time code ambiguity would be 100 years; for a digital clock in which hours and minutes up to a maximum of 11:59 are displayed, the time code ambiguity would be 12 hours.

time code resolution. The interval between two successive time code states. *Note:* Time code resolution is determined by the most rapidly changing symbol position within the time code. For example, for a digital clock that displays hours and minutes, the time code resolution would be 1 minute.

time constant. The interval required for a system or circuit to change a specified fraction from one state or condition to another. *Note 1:* The time constant is used in the expression

$$A(t) = A(0)e^{-t/a},$$

where $A(t)$ is the value of the state at time t, $A(0)$ is the value of the state at time $t = 0$, a is the time constant, and t is the time that has elapsed from the start of the exponential decay. *Note 2:* When $t = a$, $A(t)/A(0) = 1/e$, or approximately 0.37, and the system has changed about 63% toward its new value in one time constant. A system is considered to have changed its state after the elapse of three time constants, which corresponds to a 95% change in state. For example, if an electrical capacitor, having a capacitance of C farads, is discharged through a resistor, having a resistance of R ohms, the capacitor will be approximately 95% discharged after the elapse of $3RC$ seconds. *Note 3:* Time constants are expressed in seconds, such as 3.5×10^{-6} seconds, *i.e.,* 3.5 μs.

time-delay distortion. *Synonym* delay distortion.

time-derived channel. *See* time-division multiplexing.

time diversity. Transmission in which signals representing the same information are sent over the same channel at different times. *Note:* Time diversity is often used over systems subject to burst error conditions, and at intervals adjusted to be longer than an error burst.

time division. *See* time-division multiplexing.

time-division multiple access (TDMA). A communications technique that uses a common channel (multipoint or broadcast) for communications among multiple users by allocating unique time slots to different users. *Note:* TDMA is used extensively in satellite systems, local area networks, physical security systems, and combat-net radio systems.

time-division multiplexing (TDM). Digital multiplexing in which two or more apparently simultaneous channels are derived from a given frequency spectrum, *i.e.*, bit stream, by interleaving pulses representing bits from different channels. *Note:* Successive pulses represent bits from successive channels, *e.g.*, voice channels in a T1 system.

time-division switching. Switching of time-division multiplexed (TDM) channels by shifting bits between time slots in a TDM frame.

time-domain reflectometer (TDR). An electronic instrument used to characterize and locate faults in metallic cables (*e.g.*, twisted pair, coax). *Note 1:* A TDR transmits a fast rise time pulse along the conductor. The resulting reflected pulse is measured at the input as a function of time and displayed on the instrument or plotted, as a function of cable length. *Note 2:* A TDR may be used to verify cable impedance characteristics, splice and connector location and associated losses, and estimate cable lengths.

time-gated direct-sequence spread spectrum. Direct-sequence spread spectrum where the transmitter is on only for a short fraction of a time interval. The on-time can be periodic or random within a time interval.

time guard band. A time interval left vacant on a channel to provide a margin of safety against intersymbol interference in the time domain between sequential operations, such as detection, integration, differentiation, transmission, encoding, decoding, or switching.

time instability. The fluctuation of the time interval error caused by the instability of a real clock.

time interval error (TIE). The time difference between a real clock and an ideal uniform time scale, after a time interval following perfect synchronization between the clock and the scale.

time jitter. Short-term variation or instability in the duration of a specified time interval.

timeliness. *See* responsiveness.

time marker. A reference signal, often repeated periodically, enabling the correlation of specific events with a time scale, such as for establishing synchronization.

time of occurrence. The date of an event, *i.e.*, the instant an event occurs, with reference to a specified time scale.

time-out. 1. A network parameter related to an enforced event designed to occur at the conclusion of a predetermined elapsed time. 2. A specified period of time that will be allowed to elapse in a system before a specified event is to take place, unless another specified event occurs first; in either case, the period is terminated when either event takes place. *Note:* A time-out condition can be canceled by the receipt of an appropriate time-out cancellation signal. 3. An event that occurs at the end of a predetermined period of time that began at the occurrence of another specified event. The time-out can be prevented by an appropriate signal.

time scale. 1. A time measuring system defined to relate the passage of temporal events since a selected epoch. *Note:* The internationally recognized time interval is the second. Time scales are graduated in intervals such as seconds, minutes, hours, days, and years, and in fractions of a second, such as milliseconds, nanoseconds, and picoseconds. 2. Time coordinates placed on the abscissa (x-axis) of Cartesian-coordinate graphs used for depicting waveforms and similar phenomena.

time scale factor. A multiplier used to transform the real time of occurrence of an event or a problem into system time, such as that of a telecommunications system or a computer.

time-sharing. 1. The interleaving of two or more independent processes on one functional unit. 2. Pertaining to the interleaved use of computer time that enables two or more users to execute programs concurrently.

time slot. 1. Period of time during which certain activities are governed by specific regulations. 2. A time interval that can be recognized and uniquely defined.

time standard. A stable device that emits signals at equal intervals such that their count may be used as a clock.

time tick. A time mark output of a clock system.

timing extraction. *Synonym* timing recovery.

timing recovery. The derivation of a timing signal from a received signal. *Synonym* timing extraction.

timing signal. 1. The output of a clock. 2. A signal used to synchronize interconnected equipment.

timing tracking accuracy. A measure of the ability of a timing synchronization system to minimize the clock difference between a master clock and any slaved clock.

T-interface. For basic rate access in an Integrated Services Digital Network (ISDN) environment, a user-to-network interface reference point that (a) is characterized by a four-wire, 144-kb/s (2B+D) user rate, (b) accommodates the link access and transport layer function in the ISDN architecture, (c) is located at the user premises, (d) is distance sensitive to the servicing network terminating equipment, and (e) functions in a manner analogous to that of the Channel Service Units (CSUs) and the Data Service Units (DSUs).

T junction. *See* series T junction.

TLM. *Abbreviation for* telemetry.

TLP. *Abbreviation for* transmission level point.

TM. *Abbreviation for* transverse magnetic. *See* transverse magnetic mode.

TOD. *Abbreviation for* time of day. *See* time of occurrence.

token. In certain local-area-network protocols, a group of bits that serves as a symbol of authority, is passed among data stations, and is used to indicate the station that is temporarily in control of the transmission medium.

token-bus network. A bus network in which a token passing procedure is used.

token passing. A network access procedure in which a token passes from station to station and the only station allowed to transmit information is the station with the token.

token ring adapter. A network interface card (NIC) designed to attach a client workstation to a token ring computer network and operate as a token-passing interface.

token-ring network. *See* network topology.

tolerance. The permissible range of variation of some characteristic from its nominal value.

tolerance field. 1. The region between two curves, such as circles or rectangles, used to specify the tolerance on component size and geometry. 2. Pertaining to the cross section of an optical fiber, when used to specify the respective diameters and ovalities of, and concentricity error between, the core and cladding; two concentric annular regions which define the core-cladding boundary and the cladding outer boundary. *Note:* Dimensions are usually expressed in micrometers (μm). The larger annular region is defined by concentric circles of diameter $[D_C+\Delta D_C]$ and $[D_C - \Delta D_C]$, where D_C is the nominal diameter of the cladding and ΔD_C is the cladding diameter tolerance. The smaller annular region is defined by concentric circles of diameter $[D_c+\Delta D_c]$ and $[D_c - \Delta D_c]$, where D_c is the nominal diameter of the core and ΔD_c is the core diameter tolerance. When the core and cladding boundaries of the cross section of the fiber in question fall entirely within their respective defined areas, the fiber meets the specification. 3. Of the cross section of a given optical fiber, when used to characterize the respective diameters and ovalities of the core and cladding, and the concentricity error between the core and cladding; two such pairs of concentric circles, the concentric pairs not necessarily being concentric with one another. *Note 1:* One pair of concentric circles characterizes the core, and the other pair, the cladding. The cladding *ovality* is characterized by the smallest circle that circumscribes its cross section, and the largest circle that fits within its cross section. (The cross section is assumed, to a first approximation, to be elliptical in shape, so these defining circles will be concentric.) The core cross section is characterized by an analogous pair of circles, also concentric with one another, but not necessarily with those defining the cladding cross section. *Note 2:* The distance between the centers of the two concentric pairs (core pair and cladding pair) defines the offset between the core and cladding (the *Acore-cladding offset,*@also called the *Aconcentricity error*@). The width of the annulus defined by the cladding circles determines the ovality of the cladding, and the width of the annulus defined by the core determines the ovality of the core.

toll call. *See* long-distance call.

toll diversion. A system service feature by which users are denied the ability to place toll calls without the assistance of an attendant.

toll office. A central office used primarily for supervising and switching toll traffic.

toll quality. The voice quality resulting from the use of a nominal 4-kHz telephone channel. *Note:* Toll quality may be quantized in terms of a specified bit error ratio.

toll restriction. *See* classmark.

toll switching trunk. A trunk connecting one or more end offices to a toll center as the first stage of concentration for intertoll traffic. *Note:* Operator assistance or participation may be an optional function.

In U.S. common carrier telephony service, a toll center designated *"Class 4C"* is an office where assistance in completing incoming calls is provided in addition to other traffic; a toll center designated *"Class 4P"* is an office where operators handle only outbound calls, or where switching is performed without operator assistance.

T1 (carrier). *See* T-carrier.

T1C (carrier). *See* T-carrier.

T2 (carrier). *See* T-carrier.

T3 (carrier). *See* T-carrier.

T4 (carrier). *See* T-carrier.

T5 (carrier). *See* T-carrier.

tone diversity. In a voice frequency telegraph (VFTG) transmission system, the use of two channels to carry the same information. *Note:* Tone diversity is usually achieved by twinning the channels of a 16-channel VFTG to obtain 8 channels with dual diversity.

tone signaling. *See* dual-tone multifrequency signaling.

tool. *Synonym* utility program.

topography. The specification and arrangement in physical locations of actual communication and information system components which implement the topology.

topology. *See* network topology.

torn-tape relay. An antiquated tape relay system in which the perforated tape is manually transferred by an operator to the appropriate outgoing transmitter.

total channel noise. The sum of random noise, intermodulation noise, and crosstalk. *Note:* Total channel noise does not include impulse noise because different techniques are required for its measurement.

total harmonic distortion (THD). Of a signal, the ratio of (a) the sum of the powers of all harmonic frequencies above the fundamental frequency to (b) the power of the fundamental frequency. *Note 1:* The THD is usually expressed in dB. *Note 2:* Measurements for calculating the THD are made at the output of a device under specified conditions.

total internal reflection. The reflection that occurs when light, in a higher refractive-index medium, strikes an interface, with a medium with a lower refractive index, at an angle of incidence (with respect to the normal) greater than the critical angle. *See* Snell's law (Note 3).

total line length. In facsimile, the spot speed divided by the scanning line frequency. *Note:* The total line length may be greater than the length of the available line.

touch panel. *See* touch-sensitive.

touch screen. *See* touch-sensitive.

touch-sensitive. Pertaining to a device that allows a user to interact with a computer system by touching an area on the surface of the device with a finger, pencil, or other object; for example, a touch-sensitive keypad or screen.

trace packet. In a packet-switching network, a unique packet that causes a report of each stage of its progress to be sent to the network control center from each visited system element.

trace program. A computer program that performs a check on another computer program by exhibiting the sequence in which the instructions are executed and usually the results of executing the instructions.

track. On a data medium, a path associated with a single read/write head position as data move past the head.

trackball. A ball that can be rotated about its center and that is used as an input device, *e.g.*, to position a cursor. *Synonym* control ball.

track density. The number of tracks per unit length, measured in a direction perpendicular to the direction in which the tracks are read.

tracking error. The deviation of a dependent variable with respect to a reference function.

tracking mode. An operational mode during which a system is operating within specified movement limits relative to a reference.

tracking phase. *See* tracking mode.

traffic. 1. The information moved over a communication channel. 2. A quantitative measurement of the total messages and their length, expressed in CCS or other units, during a specified period of time.

traffic analysis. 1. In a communications system, the analysis of traffic rates, volumes, densities, capacities, and patterns specifically for system performance improvement. 2. [The] study of communications characteristics external to the text. 3. The analysis of the communications-electronic environment for use in the design, development, and operation of new communications systems.

traffic capacity. The maximum traffic per unit of time that a given telecommunications system, subsystem, or device can carry under specified conditions.

traffic encryption key (TEK). [A] key used to encrypt plain text or to superencrypt previously encrypted text and/or to decrypt cipher text.

traffic engineering. The determination of the numbers and kinds of circuits and quantities of related terminating and switching equipment required to meet anticipated traffic loads throughout a communications system.

traffic-flow security. 1. The protection resulting from features, inherent in some cryptoequipment, that conceal the presence of valid messages on a communications circuit; normally achieved by causing the circuit to appear busy at all times. [After JP1] 2. Measures used to conceal the presence of valid messages in an on-line cryptosystem or secure communications system. *Note:* Encryption of sending and receiving addresses and causing the circuit to appear busy at all times by sending dummy traffic are two methods of traffic-flow security. A more common method is to send a continuous encrypted signal, whether or not traffic is being transmitted.

traffic intensity. A measure of the average occupancy of a facility during a specified period of time, normally a busy hour, measured in traffic units (erlangs) and defined as the ratio of the time during which a facility is occupied (continuously or cumulatively) to the time this facility is available for occupancy. *Note:* A traffic intensity of one traffic unit (one erlang) means continuous occupancy of a facility during the time period under consideration, regardless of whether or not information is transmitted. *Synonym* call intensity.

traffic load. The total traffic carried by a trunk or trunk group during a specified time interval.

traffic monitor. In a communications network, a service feature that provides basic data on the amount and type of traffic handled by the network.

traffic overflow. 1. That condition wherein the traffic offered to a portion of a communication system exceeds its capacity and the excess may be blocked or may be provided with alternate routing. 2. The excess traffic itself.

traffic register. *See* register.

traffic service position system (TSPS). A stored program electronic system associated with one or more toll switching systems which provides centralized traffic service position functions for several local offices at one location. [47CFR part 67, Appendix.]

traffic unit. *Synonym* erlang.

traffic usage recorder. A device for measuring and recording the amount of telephone traffic carried by a group, or several groups, of switches or trunks.

transceiver. 1. A device that performs, within one chassis, both transmitting and receiving functions. 2. In military communications, the combination of transmitting and receiving equipment that (a) is in a common housing, (b) usually is designed for portable or mobile use, (c) uses common circuit components for both transmitting and receiving, and (d) provides half-duplex operation.

transcoding. The direct digital-to-digital conversion from one encoding scheme, such as voice LPC-10, to a different encoding scheme without returning the signals to analog form. *Note:* The transcoded signals, *i.e.,* the digital representations of analog signals may be any digital representation of any analog signal, such as voice, facsimile, or quasi-analog signals.

transducer. A device for converting energy from one form to another for the purpose of measurement of a physical quantity or for information transfer.

TRANSEC. *Abbreviation for* transmission security. *See* communications security.

transfer. To send information from one location and to receive it at another.

transfer characteristics. Those intrinsic parameters of a system, subsystem, or equipment which, when applied to the input of the system, subsystem, or equipment, will fully describe its output.

transfer function. 1. A mathematical statement that describes the transfer characteristics of a system, subsystem, or equipment. 2. The relationship between the input and the output of a system, subsystem, or equipment in terms of the transfer characteristics. *Note 1:* When the transfer function operates on the input, the output is obtained. Given any two of these three entities, the third can be obtained. *Note 2:* Examples of simple transfer functions are voltage gains, reflection coefficients, transmission coefficients, and efficiency ratios. An example of a complex transfer function is envelope delay distortion. *Note 3:* For a negative feedback circuit, the transfer function, *T*, is given by

$$T = \frac{e_0}{e_i} = \frac{G}{1 + GH},$$

where e_0 is the output, e_i is the input, *G* is the forward gain, and *H* is the backward gain, *i.e.,* the fraction of the output that is fed back and combined with the input in a subtracter. 3. Of an optical fiber, the complex mathematical function that expresses the ratio of the variation, as a function of modulation frequency, of the instantaneous power of the optical signal at the output of the fiber, to the instantaneous power of the optical signal that is launched into the fiber. *Note:* The optical detectors used in communication applications are square-law devices. Their output current is proportional to the input optical power. Because electrical power is proportional to current, when the optical power input drops by one-half (3 dB), the electrical power at the output of the detector drops by three-quarters (6 dB).

transfer mode. In an integrated services digital network, (ISDN), a method of transmitting, multiplexing, and switching.

transfer rate. *See* data transfer rate.

transient. *See* dynamic variation.

transit delay. Between two given points in an integrated services digital network (ISDN), the time between the moment that the first bit of a data unit, such as a frame or block, passes the first given point and the moment that bit passes the second given point, plus the transmission time of the data unit.

transition. In a signal, the changing from one significant condition to another. *Note:* Examples of transitions are the changing from one voltage level to another in a data stream, the shifting from one phase position to another in phase-shift keying, and the translation from one frequency to another in frequency-shift keying.

transition zone. *Synonym* intermediate-field region.

transit network identification. A network service feature that specifies the sequence of networks used to establish or partially establish a virtual circuit.

transit time. *Synonym* phase delay.

translating program. *Synonym* translator (def. #2).

translator. 1. A device that converts information from one system of representation into equivalent information in another system of representation. *Note:* An example of a translator in telephony is the device that converts dialed digits into call-routing information. 2. A computer program that translates from one language into another language and in particular from one programming language into another programming language. *Synonym* translating program. 3. In FM and TV broadcasting, a repeater station that receives a primary station's signal, amplifies it, shifts it in frequency, and rebroadcasts it. 4. A device that converts one frequency to another.

transliterate. To convert the characters of one alphabet to the corresponding characters of another alphabet.

transmission. 1. The dispatching, for reception elsewhere, of a signal, message, or other form of information. 2. The propagation of a signal, message, or other form of information by any means, such as by telegraph, telephone, radio, television, or facsimile via any medium, such as wire, coaxial cable, microwave, optical fiber, or radio frequency. 3. In communications systems, a series of data units, such as blocks, messages, or frames. 4. The transfer of electrical power from one location to another via conductors.

transmission block. 1. A group of bits or characters transmitted as a unit and usually containing an encoding procedure for error control purposes. 2. In data transmission, a group of records sent, processed, or recorded as a unit. *Note:* A transmission block is usually terminated by an end-of-block character (EOB), end-of-transmission-block character (ETB), or end-of-text character (EOT or ETX).

transmission channel. *See* channel.

transmission coefficient. 1. The ratio of the transmitted field strength to the incident field strength of an electromagnetic wave when it is incident upon an interface surface between media with two different refractive indices. 2. In a transmission line, the ratio of the amplitude of the complex transmitted wave to that of the incident wave at a discontinuity in the line. 3. The probability that a portion of a communications system, such as a line, circuit, channel or trunk, will meet specified performance criteria. *Note:* The value of the transmission coefficient is inversely related to the quality of the line, circuit, channel or trunk.

transmission control character. *See* control character.

transmission control protocol. A network protocol that controls host-to-host transmissions over packet-switched communication networks.

transmission frame. A data structure, beginning and ending with delimiters, that consists of fields predetermined by a protocol for the transmission of user data and control data.

transmission level. At a specified point in a telecommunications system, the power that is measured when a standard test signal, *e.g.,* 0 dBm or -16 dBm at 1000 Hz, is transmitted from a corresponding reference point. *Note:* The transmission level is usually expressed in dBm.

transmission level point (TLP). In a telecommunications system, a test point, *i.e.,* a point where a signal may be inserted or measured, and for which the nominal power of a test signal is specified. *Note 1:* In practice, the abbreviation, TLP, is usually used, and it is modified by the nominal level for the point in question. For example, where the nominal level is 0 dBm, the expression 0 dBm TLP, or simply, 0TLP, is used. Where the nominal level is -16 dBm, the expression -16 dBm TLP, or -16TLP, is used. *Note 2:* The nominal transmission level at a specified TLP is a function of system design and is an expression of the design gain or loss. *Note 3:* Voice-channel transmission levels, *i.e.,* TLPs, are usually specified for a frequency of approximately 1000 Hz. *Note 4:* The TLP at a point at which an end instrument, *e.g.*, a telephone set, is connected is usually specified as 0 dBm.

transmission line. The material medium or structure that forms all or part of a path from one place to another for directing the transmission of energy, such as electric currents, magnetic fields, acoustic waves, or electromagnetic waves. *Note:* Examples of transmission lines include wires, optical fibers, coaxial cables, rectangular closed waveguides, and dielectric slabs.

transmission loss. The decrease in power that occurs during transmission from one point to another. *Note:* Transmission loss is usually expressed in dB.

transmission medium. Any material substance, such as fiber-optic cable, twisted-wire pair, coaxial cable, dielectric-slab waveguide, water, and air, that can be used for the propagation of signals, usually in the form of modulated radio, light, or acoustic waves, from one point to another. *Note:* By extension, free space can also be considered a transmission medium for electromagnetic waves, although it is not a material medium.

transmission security. *See* communications security.

transmission security key (TSK). [A] key that is used in the control of transmission security processes, such as frequency hopping and spread spectrum.

transmission service channel. In video systems, the one-way transmission path between two designated points.

transmission system. Part of a communication system organized to accomplish the transfer of information from one point to one or more other points by means of signals. *Note:* Examples of NATO-owned transmission systems are SATCOM, ACE HIGH and CIP-67.

transmission time. In facsimile, the interval between the start of picture signals and the detection of the end-of-message signal by the receiver for a single document.

transmission window. *Synonym* spectral window. *See* window.

transmissivity. *Obsolete. See* transmittance.

transmit-after-receive time delay. The time interval from removal of rf energy at the local receiver input until the local transmitter is automatically keyed on and the transmitted rf signal amplitude has increased to 90% of its steady-state value. *An Exception:* High-frequency (HF) transceiver equipment is normally not designed with an interlock between receiver squelch and transmitter on-off key. The transmitter can be keyed on at any time, independent of whether or not a signal is being received at the receiver input.

transmit flow control. In data communications systems, control of the rate at which data are transmitted from a terminal so that the data can be received by another terminal. *Note 1:* Transmit flow control may occur between data terminal equipment (DTE) and a switching center, via data circuit-terminating equipment (DCE), or between two DTEs. The transmission rate may be controlled because of network or DTE requirements. *Note 2:* Transmit flow control can occur independently in the two directions of data transfer, thus permitting the transfer rates in one direction to be different from the transfer rates in the other direction.

transmittance. The ratio of the transmitted power to the incident power. *Note 1:* In optics, transmittance is usually expressed as optical density or in percent. *Note 2:* Transmittance was formerly called *"transmission."*

transmitter attack-time delay. The interval from the instant a transmitter is keyed-on to the instant the transmitted radio frequency (rf) signal amplitude has increased to a specified level, usually 90% of its key-on steady-state value. *Note:* The transmitter attack-time delay excludes the time required for automatic antenna tuning.

transmitter central wavelength range (λ_{tmax} - λ_{tmin}). In optical communication, the total allowed range of transmitter central wavelengths caused by the combined worst-case variations due to manufacturing, temperature, aging, and any other significant factors.

transmitter power output rating. The power output of a radio transmitter under stated conditions of operation and measurement. *Note:* Power output ratings may be made against a number of criteria, *e.g.*, peak envelope power, peak power, mean power, carrier power, noise power, or stated intermodulation level.

transmitter release-time delay. The interval from the instant a transmitter is keyed-off to the instant the transmitted radio frequency (rf) signal amplitude has decreased to a specified level, usually 10% of its key-on steady-state value.

transmultiplexer. Equipment that transforms signals derived from frequency-division multiplex equipment, such as group or supergroups, to time-division-multiplexed signals having the same structure as those derived from PCM multiplex equipment, such as primary or secondary PCM multiplex signals, and vice versa.

transparency. 1. The property of an entity that allows another entity to pass thorough it without altering either of the entities. 2. In telecommunications, the property that allows a transmission system or channel to accept, at its input, unmodified user information, and deliver corresponding user information at its output, unchanged in form or information content. *Note:* The user information may be changed internally within the transmission system, but it is restored to its original form prior to the output without the involvement of the user. 3. The quality of a data communications system or device that uses a bit-oriented link protocol that does not depend on the bit sequence structure used by the data source. 4. An image fixed on a clear base by means of a photographic printing, chemical, or other process, especially adaptable for viewing by transmitted light.

transparent interface. An interface that allows the connection and operation of a system, subsystem, or equipment with another without modification of system characteristics or operational procedures on either side of the interface.

transparent network. *See* transparency (def. #2).

transponder. 1. An automatic device that receives, amplifies, and retransmits a signal on a different frequency. 2. An automatic device that transmits a predetermined message in response to a predefined received signal. *Note:* An example of transponders is in identification-friend-or-foe systems and air-traffic-control secondary radar (beacon radar) systems. 3. A receiver-transmitter that will generate a reply signal upon proper interrogation.

transportability. 1. In communications, the quality of equipment, devices, systems, and associated hardware that permits their being moved from one location to another to interconnect with locally available complementary equipment, devices, systems, associated hardware, or other complementary facilities. *Note:* Transportability implies the use of standardized components, such as standardized plugs and transmission media. 2. The capability of material to be moved by towing, self-propulsion, or carrier via any means, such as railways, highways, waterways, pipelines, oceans, and airways.

transportable station. A station which is transferred to various fixed locations but is not intended to be used while in motion.

Transport Layer. *See* Open Systems Interconnection–Reference Model.

transposition. 1. In data transmission, a transmission defect in which, during one character period, one or more signal elements are changed from one significant condition to the other, and an equal number of elements are changed in the opposite sense. 2. In outside plant construction, an interchange of spatial positions of the several conductors of a cable between successive concatenated sections. *Note:* Transposition is usually used to minimize inductive coupling and thus reduce interference in communications circuits.

transverse electric and magnetic (TEM) mode. A mode whose electric and magnetic field vectors are both normal to the direction of propagation. *Note:* The TEM mode is the most useful mode in a coaxial cable.

transverse electric (TE) mode. A mode whose electric field vector is normal to the direction of propagation. *Note:* TE modes may be useful modes in waveguides. In an optical fiber, TE and TM modes correspond to meridional rays.

transverse magnetic (TM) mode. A mode whose magnetic field vector is normal to the direction of propagation. *Note:* TM modes may be useful in waveguides. In an optical fiber, TE and TM modes correspond to meridional rays.

transverse offset loss. *Synonym* lateral offset loss.

transverse parity check. A parity check performed on a group of binary digits recorded on parallel tracks of a data medium, such as a magnetic disk, tape, drum, or card.

transverse redundancy check (TRC). In synchronized parallel bit streams, a redundancy check (a) that is based on the formation of a block check following preset rules, (b) in which the check-formation rule applied to blocks is also applied to characters, and (c) in which the check is made on parallel bit patterns. *Note 1:* When the TRC is based on a parity bit applied to each character and block, the TRC can only detect, with limited certainty, whether or not there is an error. It cannot correct the error. Detection cannot be guaranteed because an even number of errors in the same character or block will escape detection, regardless of whether odd or even parity is used. *Note 2:* Two-dimensional arrays of bits may be used to represent characters or blocks in synchronized parallel data streams. When TRC is combined with longitudinal redundancy checking (LRC), individual erroneous bits can be corrected. *Synonym* vertical redundancy check.

transverse resolution. In a facsimile receiver, the dimension that (a) is perpendicular to a scanning line and (b) is the smallest recognizable detail of the image produced by the shortest signal capable of actuating the facsimile receiver under specified conditions.

trapped electromagnetic wave. An electromagnetic wave that enters a layer of material that is surrounded on both sides by a layer of material of a lesser refractive index such that, if the wave is traveling parallel or nearly parallel to the surfaces of the layers and hence the incident angles with the surfaces are greater than the critical angle, *i.e.*, the angles are grazing with the surface, total internal reflection will occur on both sides and hence trap the wave. *Note:* Dielectric slabs, optical fibers, and layers of air can serve as an electromagnetic wave trap, thus confining the wave to a given direction of propagation and to a given point.

trapped mode. *Synonym* bound mode.

trapped ray. *Synonym* guided ray.

traveling wave. A wave that (a) propagates in a transmission medium, (b) has a velocity determined by the launching conditions and the physical properties of the medium, and (c) may be a longitudinal or transverse wave. *Note 1:* For the purposes of this definition, free space may be considered a medium, although it is not a physical medium. *Note 2:* A traveling wave is not a wave that is reduced to a standing wave by reflections from a distant boundary. *Note 3:* Examples of traveling waves are radio waves propagating in free space, lightwaves propagating in optical fibers, water waves on the surface of the ocean, and seismic waves.

tree network. *See* network topology.

tree search. In a tree structure, a search in which it is possible to decide, at each step, which part of the tree may be rejected without a further search.

tree structure. A hierarchical organization in which a given node is considered to be an ancestor of all the lower level nodes to which the given node is connected. *Note 1:* The root node, *i.e.,* the base node, is an ancestor of all the other nodes. *Note 2:* In a tree structure there is one and only one path from any point to any other point.

tree topology. *See* network topology.

T reference point. In Integrated Services Digital Networks (ISDN), the conceptual point dividing NT2 and NT1 functional groupings in a particular ISDN arrangement.

tremendously high frequency (THF). Frequencies from 300 GHz to 3000 GHz.

tributary office. A local office, located outside the exchange in which a toll center is located, that has a different rate center from its toll center.

tributary station. 1. In a data network, a station other than the control station. 2. On a multipoint connection or a point-to-point connection using basic mode link control, any data station other than the control station.

trim effect. In a crystal oscillator, the degradation of frequency-vs.-temperature stability, and marked frequency offset, resulting from frequency adjustment which produces a rotation or distortion, or both, of the inherent frequency-vs.-temperature characteristic.

triple precision. Characterized by the use of three computer words to represent a number in accordance with required precision.

triplet. A byte composed of three bits. *Synonym* three-bit byte.

TRI-TAC. *Acronym for* **tri-services tactical.** *See* tactical communications.

TRI-TAC equipment. Equipment that (a) accommodates the transition from current manual and analog systems to fully automated digital systems and (b) provides for message switching, voice communications circuit switching, and the use of secure voice terminals, digital facsimile systems, and user digital voice terminals.

troposcatter. *Synonym* tropospheric scatter.

troposphere. 1. The lower layers of atmosphere, in which the change of temperature with height is relatively large. It is the region where clouds form, convection is active, and mixing is continuous and more or less complete. 2. The layer of the Earth's atmosphere, between the surface and the stratosphere, in which temperature decreases with altitude and which contains approximately 80% of the total air mass. *Note:* The thickness of the troposphere varies with season and latitude. It is usually 16 km to 18 km thick over tropical regions, and less than 10 km thick over the poles.

tropospheric duct. *See* atmospheric duct.

tropospheric scatter. 1. The propagation of radio waves by scattering as a result of irregularities or discontinuities in the physical properties of the troposphere. 2. A method of transhorizon communications using frequencies from approximately 350 MHz to approximately 8400 MHz. *Note:* The propagation mechanism is still not fully understood, though it includes several distinguishable but changeable mechanisms such as propagation by means of random reflections and scattering from irregularities in the dielectric gradient density of the troposphere, smooth-Earth diffraction, and diffraction over isolated obstacles (knife-edge diffraction). *Synonym* **troposcatter.**

tropospheric wave. A radio wave that is propagated by reflection from a place of abrupt change in the dielectric constant, or its gradient, in the troposphere. *Note:* In some cases, a ground wave may be so altered that new components appear to arise from reflection in regions of rapidly changing dielectric constant. When these components are distinguishable from the other components, they are called *"tropospheric waves."*

true power. *Synonym* effective power.

truncated binary exponential backoff. In carrier sense multiple access with collision avoidance (CSMA/CA) networks and in carrier sense multiple access with collision detection (CSMA/CD) networks, the algorithm used to schedule retransmission after a collision such that the retransmission is delayed by an amount of time derived from the slot time and the number of attempts to retransmit.

truncation. The deletion or omission of a leading or a trailing portion of a string in accordance with specified criteria.

truncation error. In the representation of a number, the error introduced when one or more digits are dropped.

trunk. 1. In a communications network, a single transmission channel between two points that are switching centers or nodes, or both. 2. [A] circuit between switchboards or other switching equipment, as distinguished from circuits which extend between central office switching equipment and information origination/termination equipment. *Note:* Trunks may be used to interconnect switches, such as major, minor, public and private switches, to form networks.

trunk encryption device (TED). A bulk encryption device used to provide secure communications over a wideband digital transmission link. *Note:* A TED is usually located between the output of a trunk group multiplexer and a wideband radio or cable facility.

trunk group. Two or more trunks of the same type between two given points.

trunk group multiplexer (TGM). A time-division multiplexer that combines individual digital trunk groups into a higher rate bit stream for transmission over wideband digital communications links.

trunk hunting. *See* hunting (def. #1).

trusted computer system (TCS). [An] AIS that employs sufficient hardware and software assurance measures to allow simultaneous processing of a range of classified or sensitive information.

trusted computing base (TCB). [The] totality of protection mechanisms within a computer system, including hardware, firmware, and software, the combination of which is responsible for enforcing a security policy. *Note:* The ability of a trusted computing base to enforce correctly a unified security policy depends on the correctness of the mechanisms within the trusted computing base, the protection of those mechanisms to ensure their correctness, and the correct input of parameters related to the security policy.

truth table. 1. An operation table for a logic operation. 2. A table that describes a logic function by listing all possible combinations of input values and indicating, for each combination, the output value.

TSK. *Abbreviation for* transmission security key.

TSP. *Abbreviation for* Telecommunications Service Priority.

TSPS. *Abbreviation for* traffic service position system.

TSP system. *See* Telecommunications Service Priority system.

TTTN. *Abbreviation for* tandem tie trunk network.

TTY. *Abbreviation for* teletypewriter.

TTY/TDD. A unique telecommunication device for the deaf, using TTY principles.

tuning. Adjusting the parameters and components of a circuit so that it resonates at a particular frequency or so that the current or voltage is either maximized or minimized at a specific point in the circuit. *Note:* Tuning is usually accomplished by adjusting the capacitance or the inductance, or both, of elements that are connected to or in the circuit.

tunneling mode. *Synonym* leaky mode.

tunneling ray. *Synonym* leaky ray.

Turing machine. A mathematical model of a device that changes its internal state and reads from, writes on, and moves a potentially infinite tape, all in accordance with its present state, thereby constituting a model for computer-like behavior.

turnaround time. In a half-duplex circuit, the time required to reverse the direction of transmission from transmit to receive or vice versa.

turnkey. Pertaining to a procurement process that (a) includes contractual actions at least through the system, subsystem, or equipment installation phase and (b) may include follow-on contractual actions, such as testing, training, logistical, and operational support. *Note:* Precise definition of the types of allowable contractual features are contained in the Federal Acquisition Regulations (FAR).

twin cable. A cable composed of two parallel conductors separated from each other by a ribbon-like insulator or encased by a foam insulator. *Synonym* twin-lead.

twin-lead. *Synonym* twin cable.

twinplex. A frequency-shift-keyed (FSK) carrier telegraphy system in which four unique tones, *i.e.,* two pairs of tones, are transmitted over a single transmission channel, such as one twisted pair. *Note:* One tone of each pair represents a *"mark"* and the other a *"space."*

twin sideband transmission. *See* independent-sideband transmission.

twist. In telephony, a change, as a function of temperature, in the shape of the frequency-vs.-attenuation response curve, *i.e.,* characteristic, of a transmission line.

twisted pair cable. *See* paired cable.

two-out-of-five code. A binary-coded decimal notation in which (a) each decimal digit is represented by a binary numeral consisting of five binary digits of which two are of one kind, called *"ones,"* and three are of the other kind, called *"zeros"* and (b) the usual weights assigned to the digit positions are 0-1-2-3-6, except that *"zero"* is represented as 01100.

two-pilot regulation. In frequency-division multiplexed (FDM) systems, the use of two pilot frequencies within a band so that the differential change in attenuation with respect to temperature, *i.e.,* twist, can be detected and compensated by a regulator.

two-sample deviation. The square root of the Allan variance.

two-sample variance. *Synonym* Allan variance.

two-source frequency keying. *Synonym* frequency-exchange signaling.

two-tone keying. In telegraphy systems, keying in which the modulating wave causes the carrier to be modulated with a single tone for the *"mark"* and modulated with a different single tone for the *"space."*

two-tone telegraph. *See* two-tone keying.

two-way alternate operation. *Synonym* half-duplex operation.

two-way simultaneous operation. *Synonym* duplex operation.

two-wire circuit. A communications circuit formed by two metallic conductors insulated from each other. *Contrast with* four-wire circuit.

TWX®. *Acronym for* teletypewriter exchange service.

TX. *Abbreviation for* transmitter, transmit.

type 1 product. [A] classified or controlled cryptographic item endorsed by the National Security Agency for securing classified and sensitive U.S. Government information, when appropriately keyed. *Note:* The term refers only to products, and not to information, key, services, or controls. Type 1 products contain classified National Security Agency algorithms. They are available to U.S. Government users, their contractors, and federally sponsored non-U.S. Government activities subject to export restrictions in accordance with International Traffic in Arms Regulation.

type 2 product. Unclassified cryptographic equipment, assembly, or component, endorsed by the National Security Agency, for use in telecommunications and automated information systems for the protection of national security information. *Note:* The term refers only to products, and not to information, key, services, or controls. Type 2 products may not be used for classified information, but contain classified National Security Agency algorithms that distinguish them from products containing the unclassified data algorithm. Type 2 products are subject to export restrictions in accordance with the International Traffic in Arms Regulation.

type 3 algorithm. [A] cryptographic algorithm that has been registered by the National Institute of Standards and Technology and has been published as a Federal Information Processing Standard for use in protecting unclassified sensitive information or commercial information.

type 4 algorithm. [An] unclassified cryptographic algorithm that has been registered by the National Institute of Standards and Technology, but is not a Federal Information Processing Standard.

UDP. *Abbreviation for* user datagram protocol. An Internet protocol for datagram service.

UHF. *Abbreviation for* ultra high frequency. *See* electromagnetic spectrum.

U interface. For basic-rate access in an Integrated Services Digital Network (ISDN) environment, a user-to-network interface reference point that is characterized by the use of a 2-wire-loop transmission system that (a) conveys information between the 4-wire user-to-network interface, *i.e.,* the S/T reference point, and the local exchange, (b) is located in the servicing central office, and (c) is not as distance sensitive as a service using a T interface.

ULF. *Abbreviation for* ultra low frequency. *See* electromagnetic spectrum.

ultra high frequency (UHF). Frequencies from 300 MHz to 3000 MHz.

ultra low frequency (ULF). Frequencies from 300 Hz to 3000 Hz.

ultraviolet (uv). The portion of the electromagnetic spectrum in which the longest wavelength is just below the visible spectrum, extending from approximately 4 nm to approximately 400 nm. *Note:* Some authorities place the lower limit of uv at values between 1 and 40 nm, 1 nm being the upper wavelength limit of x-rays. The 400-nm limit is the lowest visible wavelength, *i.e.,* the highest visible frequency, violet.

unallowable character. *Synonym* illegal character.

unavailability. A expression of the degree to which a system, subsystem, or equipment is not operable and not in a committable state at the start of a mission, when the mission is called for at an unknown, *i.e.,* random, time. *Note 1:* The conditions determining operability and committability must be specified. *Note 2:* Expressed mathematically, unavailability is 1 minus the availability. *Note 3:* Unavailability may also be expressed mathematically as the ratio of the total time a functional unit is not capable of being used during a given interval to the length of the interval, *e.g.,* if the unit is not capable of being used for 68 hours a week, the unavailability is 68/168.

unbalanced line. A transmission line, such as a coaxial cable, in which the magnitudes of the voltages on the two conductors are not equal with respect to ground.

unbalanced modulator. A modulator in which the modulation factor is different for the alternate half-cycles of the carrier. *Synonym* asymmetrical modulator.

unbalanced wire circuit. A circuit in which the two sides are inherently electrically dissimilar.

unbound mode. *Synonym* radiation mode.

unbundling. In the context of the FCC's Computer III Inquiry, the process of separating individual tariffed offerings and services that are associated with a specific element in the CEI or ONA tariff from other tariffed basic service offerings.

underflow. In computing, a condition occurring when a machine calculation produces a non-zero result that is smaller than the smallest non-zero quantity that the machine's storage unit is capable of storing or representing.

underground cable. A communication cable designed to be placed under the surface of the Earth in a duct system that isolates it from direct contact with the soil. *Contrast with* direct-buried cable.

underlap. In facsimile, a defect that occurs when the width of the scanning line is less than the scanning pitch.

undershoot. *See* **over**shoot.

undesired signal. Any signal that tends to produce degradation in the operation of equipment or systems.

undetected error rate. *Deprecated synonym for* undetected error ratio.

undetected error ratio. The ratio of the number of bits, unit elements, characters, or blocks incorrectly received and undetected, to the total number of bits, unit elements, characters, or blocks sent. *Synonyms* residual error rate, undetected error rate *[deprecated]*.

undisturbed day. A day during which neither sunspot activity nor ionospheric disturbance causes detectable interfere with radio communications.

UNICOM station. *Synonym* aeronautical advisory station.

unidirectional channel. *Synonym* one-way-only channel.

unidirectional operation. Operation in which data are transmitted from a transmitter to a receiver in only one direction.

uniform encoding. An analog-to-digital conversion process in which, except for the highest and lowest quantization steps, all of the quantization subrange values are equal. *Synonym* uniform quantizing.

uniform linear array. An antenna composed of a relatively large number of usually identical elements arranged in a single line or in a plane with uniform spacing and usually with a uniform feed system.

uniform quantizing. *Synonym* uniform encoding.

uniform-spectrum random noise. *See* white noise.

uniform time scale. A time scale made up of equal intervals.

uniform transmission line. A transmission line that has distributed electrical properties, *i.e.,* resistance, inductance, and capacitance per unit length, that are constant along the line, and in which the voltage-to-current ratio does not vary with distance along the line, if the line is terminated in its characteristic impedance. *Note 1:* Examples of uniform transmission lines are coaxial cables, twisted pairs, and single wires at constant height above ground, all of which have no changes in geometry, materials, or construction along their length. *Note 2:* In a uniform transmission line, signal attenuation is a function of the length of the line and the frequency of the signal.

unilateral control system. *Synonym* unilateral synchronization system.

unilateral synchronization system. A system of synchronization in which signals from a single location are used to synchronize clocks at one or more other locations. *Synonym* unilateral control system.

unimode fiber. *Synonym* single-mode optical fiber.

unintelligible crosstalk. Crosstalk that consists of unintelligible signals, hence from which information cannot be derived.

unintentional interference. *See* interference.

uninterruptible power supply (UPS). A device that is inserted between a primary power source, such as a commercial utility, and the primary power input of equipment to be protected, *e.g.*, a computer system, for the purpose of eliminating the effects of transient anomalies or temporary outages. *Note 1:* An UPS consists of an inverter, usually electronic, that is powered by a battery that is kept trickle-charged by rectified ac from the incoming power line fed by the utility. In the event of an interruption, the battery takes over without the loss of even a fraction of a cycle in the ac output of the UPS. The battery also provides protection against transients. The duration of the longest outage for which protection is ensured depends on the battery capacity, and to a certain degree, on the rate at which the battery is drained. *Note 2:* An UPS should not be confused with a standby generator, which may not provide protection from a momentary power interruption, or which may result in a momentary power interruption when it is switched into service, whether manually or automatically.

unipolar signal. A two-state signal where one of the states is represented by voltage or current and the other state is represented by no voltage or current. *Note:* The current flow can be in either direction.

unique key. Key held only by one crypto-equipment and its associated distribution center.

unit-distance code. An unweighted code that changes at only one digit position when going from one number to the next in a consecutive sequence of numbers. *Note 1:* Use of one of the many unit-distance codes can minimize errors at symbol transition points when converting analog quantities into digital quantities. *Note 2:* An example of a unit-distance code is the Gray code.

unit element. In the representation of a character, a signal element that has a duration equal to the unit interval.

unit impulse. A mathematical artifice consisting of an impulse of infinite amplitude and zero width, and having an area of unity. *Note:* The unit impulse is useful for the mathematical expression of the impulse response, *i.e.,* the transfer function, of a device. *Synonym* Dirac delta function.

unit interval. In isochronous transmission, the longest interval of which the theoretical durations of the significant intervals of a signal are all whole multiples.

universal personal telecommunications number. *Synonym* UPT number.

Universal Personal Telecommunications (UPT) service. A telecommunications service that provides personal mobility and service profile management. *Note 1:* UPT service involves the network capability of identifying uniquely a UPT user by means of a UPT number. *Note 2:* The general principles of UPT are given in ITU-T Recommendation F.850. *Note 3:* UPT and PCS are sometimes mistakenly assumed to be the same service concept. UPT allows complete personal mobility across multiple networks and service providers. PCS may use UPT concepts to improve subscriber mobility in allowing roaming to different service providers, but UPT and PCS are not the same service concept.

universal service. The concept of making basic local telephone service (and, in some cases, certain other telecommunications and information services) available at an affordable price to all people within a country or specified jurisdictional area.

Universal Time (UT). 1. The basis for coordinated dissemination of time signals, counted from 0000 at midnight. 2. In celestial navigation applications, the time which gives the exact rotational orientation of the Earth obtained from UTC by applying increments determined by the U.S. Naval Observatory. 3. A measure of time that conforms, within a close approximation, to the mean diurnal rotation of the Earth and serves as the basis of civil timekeeping. *Note:* Universal Time (UT1) is determined from observations of the stars, radio sources, and also from ranging observations of the Moon and artificial Earth satellites. The scale determined directly from such observations is designated Universal Time Observed (UTO); it is slightly dependent on the place of observation. When UTO is corrected for the shift in longitude of the observing station caused by polar motion, the time scale UT1 is obtained. When an accuracy better than one second is not required, Universal Time can be used to mean Coordinated Universal Time (UTC). 4. The official civil time of the United Kingdom. *Formerly called* Greenwich Mean Time. *Contrast with* Coordinated Universal Time.

UNIX™. A portable, multiuser, time-shared operating system that supports process scheduling, job control, and a programmable user interface. *Note 1:* There are many proprietary operating systems that are based on UNIX™ and are colloquially referred to as UNIX™, but are not necessarily interoperable. *Note 2:* Most UNIX™-based operating systems are POSIX compliant.

unnumbered command. In a data transmission, a command that does not contain sequence numbers in the control field.

unnumbered response. In data transmission, a response that does not contain sequence numbers in the control field.

unsuccessful call. *Synonym* unsuccessful call attempt.

unsuccessful call attempt. A call attempt that does not result in the establishment of a connection. *Synonym* unsuccessful call.

unused character. *Synonym* illegal character.

up-converter. A device that translates frequencies from lower to higher frequencies.

update. The regeneration of a display to show current status, based on changes to the previously displayed data. *Note:* An update can be accomplished upon user request or by automatic means.

uplink (U/L). 1. The portion of a communications link used for the transmission of signals from an Earth terminal to a satellite or to an airborne platform. *Note:* An uplink is the converse of a downlink. 2. Pertaining to data transmission from a data station to the headend.

upright position. *Synonym* erect position.

UPS. *Acronym for* uninterruptible power supply.

upstream. 1. The direction opposite the data flow. 2. With respect to the flow of data in a communications path: at a specified point, the direction toward which data are received earlier than at the specified point.

UPT. *See* Universal Personal Telecommunications service.

UPT access code. In universal personal telecommunications service, the code that UPT users may need to dial, when using some terminals and networks, to enter the UPT environment before executing any UPT procedures.

UPT database. In universal personal telecommunications service, a repository for information, such as a service profile, that is related to a set of UPT customers and UPT users.

UPT environment. In universal personal telecommunications service, the environment within which the UPT service facilities are offered, consisting of combinations of networks and UPT service control facilities that, when combined, enable UPT users to make use of telecommunication services offered by these networks. *Note:* To the UPT user, the UPT environment appears as a single global network that provides personal mobility. However, when utilizing telecommunication services, the UPT network user may be limited by restrictions imposed by the network, by the capabilities of the terminal and network used, or by regulatory requirements.

uptime. The time during which a functional unit is fully operational.

UPT indicator. In universal personal telecommunications, that portion (or portions) of the UPT number that identifies a call as a UPT call.

UPT number. In universal personal telecommunications, the number that uniquely identifies a UPT user and that is used to place a call to, or to forward a call to, that user. *Note:* A user may have multiple UPT numbers, *e.g.,* a business UPT number for business calls and a private UPT number for private calls. In the case of multiple numbers, each UPT number is considered, from a network vantage point, to identify a distinct UPT user, even if all such numbers identify the same person or entity. *Synonym* universal personal telecommunications number.

UPT routing address. In universal personal telecommunications, the number used by the network to direct a call according to the user's UPT service profile.

UPT service profile. In universal personal telecommunications, a record that contains all information related to a UPT user, which information is required to provide that user with UPT service such as subscriptions to basic and supplementing services and call-routing preferences. *Note:* Each UPT service profile is associated with a single UPT number. *Synonym* service profile.

UPT service profile management. In universal personal telecommunications, authorized access to and manipulation of the UPT service profile. *Note:* UPT service profile management can be performed by the UPT user, by the UPT customer (subscriber), or by the UPT service provider. *Synonym* service profile management.

UPT subscriber. In universal personal telecommunications, a person who (or entity that) obtains a UPT service from a UPT service provider on behalf of one or more UPT users.

UPT user. In universal personal telecommunications, a person who (or entity that) has access to universal personal telecommunication (UPT) services and who has been assigned a UPT number.

UPT user group. A specific set of universal personal telecommunication users.

usable line length. *See* available line.

usage. *Synonym* occupancy.

useful line. *Synonym* available line.

user. 1. A person, organization, or other entity (including a computer or computer system), that employs the services provided by a telecommunication system, or by an information processing system, for transfer of information. *Note:* A user functions as a source or final destination of user information, or both. *Synonym* subscriber. 2. [A] person or process accessing an AIS by direct connections (*e.g.*, via terminals) or indirect connections. *Note:* "Indirect connection" relates to persons who prepare input data or receive output that is not reviewed for content or classification by a responsible individual.

User Datagram Protocol (UDP). In the Internet Protocol suite, a standard, low-overhead, connectionless, host-to-host protocol that is used over packet-switched computer communications networks, and that allows an application program on one computer to send a datagram to an application program on another computer. *Note:* The main difference between UDP and TCP is that UDP provides connectionless service, whereas TCP does not.

user information. Information transferred across the functional interface between a source user and a telecommunications system for delivery to a destination user. *Note:* In telecommunications systems, user information includes user overhead information.

user information bit. A bit transferred from a source user to a telecommunications system for delivery to a destination user. *Note 1:* User information bits do not include the overhead bits originated by, or having their primary functional effect within, the telecommunications system. *Note 2:* User information bits are encoded to form channel bits.

user information block. A block that contains at least one user information bit.

user line. *Synonym* loop (def. #2).

user overhead information. *See* overhead information.

user service class. *Synonym* class of service.

UT. *Abbreviation for* Universal Time.

UTC. *Abbreviation for* Coordinated Universal Time.

UTC(i). Coordinated Universal Time (UTC), as kept by the "*i*" laboratory, where *i* is any laboratory cooperating in the determination of UTC. *Note:* In the United States, the official UTC is kept by the U.S. Naval Observatory and is referred to as UTC (USNO).

utility load. *Synonym* nonoperational load.

utility program. A computer program that is in general support of the operations and processes of a computer. *Note:* Examples of utility programs include diagnostic programs, trace programs, input routines, and programs used to perform routine tasks, *i.e.,* perform everyday tasks, such as copying data from one storage location to another. *Synonyms* service program, service routine, tool, utility routine.

utility routine. *Synonym* utility program.

uv. *Abbreviation for* ultraviolet.

validation. 1. Tests to determine whether an implemented system fulfills its requirements. 2. The checking of data for correctness or for compliance with applicable standards, rules, and conventions. 3. [The] process of applying specialized security test and evaluation procedures, tools, and equipment needed to establish acceptance for joint usage of an AIS by one or more departments or agencies and their contractors. *Note:* This action will include, as necessary, final development, evaluation, and testing, preparatory to acceptance by senior security test and evaluation staff specialists. 4. In universal personal telecommunications, the process of verifying that a user or terminal is authorized to access UPT services.

value-added carrier. A company that sells the services of a value-added network.

value-added network (VAN). A network using the communication services of other commercial carriers, using hardware and software that permit enhanced telecommunication services to be offered.

VAN. *Acronym for* value-added network.

variable length buffer. A buffer into which data may be entered at one rate and removed at another rate without changing the data sequence. *Note:* Most first-in first-out (FIFO) storage devices are variable-length buffers in that the input rate may be variable while the output rate is constant or the output rate may be variable while the input rate is constant. Various clocking and control systems are used to allow control of underflow or overflow conditions.

variable slope delta modulation. *See* **continuously variable slope delta modulation.**

******variance.** In statistics, in a population of samples, the mean of the squares of the differences between the respective samples and their mean, expressed mathematically as:

where *n* is the number of samples, *x* is the value of sample *i*, is the mean of the samples, $_i$and ó^2 is the variance. *Note:* The square root of the variance, ó, is the standard deviation.

variant. 1. One of two or more code symbols which have the same plain text equivalent. 2. One of several plain text meanings that are represented by a single code group.

variation monitor. In ac power distribution, a device for sensing deviations of any measured variable, such as voltage, current, or frequency, and capable of initiating a programmed action, such as transfer to other power sources, when programmed limits of voltage, current, frequency, or time are exceeded, or providing an alarm, or both.

vars. *Abbreviation for* volt-amperes reactive.

VC. *Abbreviation for* virtual circuit.

VDU. *Abbreviation for* **vis**ual display unit. *See* monitor.

vector processor. *Synonym* array processor.

verified off-hook. In telephone systems, a service provided by a unit that is inserted on each end of a transmission circuit for the purpose of verifying supervisory signals on the circuit. *See* automatic ringdown circuit.

vertex angle. In an optical fiber, the angle formed by the extreme bound meridional rays accepted by the fiber, or emerging from it, equal to twice the acceptance angle; the angle formed by the largest cone of light accepted by the fiber or emitted from it.

vertical redundancy check (VRC). *Synonym* transverse redundancy check.

very high frequency (VHF). Frequencies from 30 MHz to 300 MHz.

very low frequency (VLF). Frequencies from 3 kHz to 30 kHz.

vestigial sideband (VSB) transmission. Modified AM transmission in which one sideband, the carrier, and only a portion of the other sideband are transmitted.

VF. *Abbreviation for* voice frequency.

VFCT. *Abbreviation for* voice frequency carrier telegraph. *See* voice-frequency telegraph.

VFCTG. *Abbreviation for* voice-frequency carrier telegraph. *See* voice-frequency telegraph.

VF patch bay. *See* voice frequency primary patch bay.

VFTG. *Abbreviation for* voice-frequency telegraph.

VHF. *Abbreviation for* very high frequency. *See* electromagnetic spectrum.

via net loss (VNL). Pertaining to circuit performance prediction and description that allows circuit parameters to be predetermined and the circuit to be designed to meet established criteria by analyzing actual, theoretical, and calculated losses.

video. 1. An electrical signal containing timing (synchronization), luminance (intensity), and often chrominance (color) information that, when displayed on an appropriate device, gives a visual image or representation of the original image sequences. 2. Pertaining to the sections of a television system that carry television signals, either in unmodulated or modulated form. 3. Pertaining to the demodulated radar signal that is applied to a radar display device. 4. Pertaining to the bandwidth or data rate necessary for the transmission of real-time television pictures. *Note:* In practice, the baseband bandwidth required for the transmission of NTSC television pictures (not including the audio carriers) is approximately 5 MHz.

video codec. *See* codec.

videoconference. *See* video teleconference.

video display terminal. *Synonym* visual display unit. *See* monitor.

video display unit. *Synonym* visual display terminal. *See* monitor.

video frame. *See* frame (def. #6).

videophone. 1. A telephone that is coupled to an imaging device that enables the call receiver or the call originator, or both, to view one another as on television, if they so desire. 2. A military communications terminal that (a) has video teleconference capability, (b) is usually configured as a small desktop unit, designed for one operator, and (c) is a single, integrated unit.

video teleconference. 1. A teleconference that includes video communications. 2. Pertaining to a two-way electronic communications system that permits two or more persons in different locations to engage in the equivalent of face-to-face audio and video communications. *Note*: Video teleconferences may be conducted as if all of the participants are in the same room.

video teleconferencing unit (VTU). Equipment that performs video teleconference functions, such as coding and decoding of audio and video signals and multiplexing of video, audio, data, and control signals, and that usually does not include I/O devices, cryptographic devices, network interface equipment, network connections, or the communications network to which the unit is connected.

view. In satellite communications, the quality or degree of visibility of a satellite to a ground station; *i.e.*, the degree to which the satellite is sufficiently above the horizon and clear of obstructions so that it is within a clear line of sight by an Earth terminal. *Note:* A pair of satellite Earth terminals has a satellite in mutual view when both have unobstructed line-of-sight contact with the satellite simultaneously.

viewdata. A type of information-retrieval service in which a subscriber can (a) access a remote database via a common carrier channel, (b) request data, and (c) receive requested data on a video display over a separate channel. *Note:* The access, request, and reception are usually via common carrier broadcast channels. *Contrast with* teletext.

violation. *See* AMI violation.

virtual call. A call, established over a network, that uses the capabilities of either a real or virtual circuit by sharing all or any part of the resources of the circuit for the duration of the call.

virtual call capability. A service feature in which (a) a call set-up procedure and a call disengagement procedure determine the period of communication between two data terminal equipments (DTEs) in which user data are transferred by the network in the packet mode of operation, (b) end-to-end transfer control of packets within the network is required, (c) data may be delivered to the network by the call originator before the call access phase is completed, but the data are not delivered to the call receiver if the call attempt is unsuccessful, (d) the network delivers all the user data to the call receiver in the same sequence in which the data are received by the network, and (e) multi-access DTEs may have several virtual calls in progress at the same time. *Synonym* virtual call facility.

virtual call facility. *Synonym* virtual call capability.

virtual carrier frequency. In radio or carrier systems in which no carrier is transmitted, *e.g.,* single sideband or double sideband with suppressed carrier, the location in the frequency spectrum that the carrier would occupy if it were present.

virtual circuit (VC). A communications arrangement in which data from a source user may be passed to a destination user over various real circuit configurations during a single period of communication. *Note:* Virtual circuits are generally set up on a per-call basis and are disconnected when the call is terminated; however, a permanent virtual circuit can be established as an option to provide a dedicated link between two facilities. *Synonyms* logical circuit, logical route.

virtual circuit capability. A network-provided service feature in which a user is provided with a virtual circuit. *Note:* Virtual circuit capability is not necessarily limited to packet mode transmission. For example, an analog signal may be converted to a digital signal and then be routed over the network via any available route.

virtual connection. A logical connection that is made to a virtual circuit.

virtual height. The apparent height of an ionized layer, as determined from the time interval between the transmitted signal and the ionospheric echo at vertical incidence.

virtual instruction. *Synonym* teletraining.

virtual memory. In computer systems, the memory as it appears to, *i.e.,* as it is available to, the operating programs running in the central processing unit (CPU). *Note:* The virtual memory may be smaller, equal to, or larger than the real memory present in the system.

virtual network. A network that provides virtual circuits and that is established by using the facilities of a real network.

virtual path. *See* virtual circuit.

virtual reality. An interactive, computer-generated simulated environment with which users can interact using specialized peripherals such as data gloves and head-mounted computer-graphic displays.

virtual storage. The storage space that may be regarded as addressable main storage by the user of a computer system in which virtual addresses are mapped into real addresses. *Note:* The size of virtual storage is limited by the addressing scheme of the computer system and by the amount of auxiliary storage available, and not by the actual number of main storage locations.

virtual terminal (VT). In open systems, an application service that (a) allows host terminals on a multi-user network to interact with other hosts regardless of terminal type and characteristics, (b) allows remote log-on by local-area-network managers for the purpose of management, (c) allows users to access information from another host processor for transaction processing, and (d) serves as a backup facility.

virus. An unwanted program which places itself into other programs which are shared among computer systems, and replicates itself. *Note:* A virus is usually manifested by a destructive or disruptive effect on the executable program that it affects.

visible spectrum. The region of the electromagnetic spectrum that can be perceived by human vision, approximately the wavelength range of 0.4 μm to 0.7 μm.

visual display unit (VDU). *See* monitor.

vitreous silica. Glass consisting of almost pure silicon dioxide (SiO_2). *Synonym* fused silica.

VLF. *Abbreviation for* very low frequency. *See* electromagnetic spectrum.

VNL. *Abbreviation for* via net loss.

V number. *Synonym* normalized frequency (def. #1).

vocoder. *Abbreviation for* voice-coder. A device that usually consists of a speech analyzer, which converts analog speech waveforms into narrowband digital signals, and a speech synthesizer, which converts the digital signals into artificial speech sounds. *Note 1:* For COMSEC purposes, a vocoder may be used in conjunction with a key generator and a modulator-demodulator to transmit digitally encrypted speech signals over narrowband voice communications channels. These devices are used to reduce the bandwidth requirements for transmitting digitized speech signals. *Note 2:* Some analog vocoders move incoming signals from one portion of the spectrum to another portion.

vodas. *Acronym for* voice-operated device anti-sing. A device used to prevent overall voice-frequency singing in a two-way telephone circuit by ensuring that transmission can occur in only one direction at any given instant.

vogad. *Acronym for* voice-operated gain-adjusting device. A device that has a substantially constant output amplitude over a wide range of input amplitudes.

voice band. *Synonym* voice frequency.

voice coder. *See* vocoder.

voice-data signal. *See* quasi-analog signal.

voice frequency (VF). Pertaining to those frequencies within that part of the audio range that is used for the transmission of speech. *Note 1:* In telephony, the usable voice-frequency band ranges from approximately 300 Hz to 3400 Hz. *Note 2:* In telephony, the bandwidth allocated for a single voice-frequency transmission channel is usually 4 kHz, including guard bands. *Synonym* voice band.

voice frequency carrier telegraph (VFCT). *Synonym* voice-frequency telegraph.

voice-frequency (VF) channel. A channel capable of carrying analog and quasi-analog signals.

voice frequency (VF) primary patch bay. A patching facility that provides the first appearance of local-user VF circuits in the technical control facility (TCF). *Note:* The VF primary patch bay provides patching, monitoring, and testing for all VF circuits. Signals will have various levels and signaling schemes depending on the user terminal equipment.

voice-frequency telegraph (VFTG). A method of multiplexing one or more dc telegraph channels onto a nominal 4-kHz voice frequency channel. *Synonym* voice frequency carrier telegraph.

voice grade. In the public regulated telecommunications services, a service grade that is described in part 68, Title 47 of the *Code of Federal Regulations [CFR]. Note:* Voice-grade service does not imply any specific signaling or supervisory scheme.

voice-operated device anti-sing. *See* vodas.

voice-operated gain-adjusting device. *See* vogad.

voice operated relay circuit. *Synonym* vox.

voice operated transmit. *Synonym* vox.

voice-plus circuit. *Synonym* composited circuit.

volatile storage. A storage device in which the contents are lost when power is removed.

volatility. *See* data volatility.

voltage standing wave ratio (VSWR). In a transmission line, the ratio of maximum to minimum voltage in a standing wave pattern. *Note:* The VSWR is a measure of impedance mismatch between the transmission line and its load. The higher the VSWR, the greater the mismatch. The minimum VSWR, *i.e.,* that which corresponds to a perfect impedance match, is unity.

volt-amperes reactive (vars). In alternating-current power transmission and distribution, the product of the rms voltage and amperage, *i.e.,* the apparent power, multiplied by the sine of the phase angle between the voltage and the current. *Note 1:* Vars represents the power not consumed by a reactive load, *i.e.,* when there is a phase difference between the applied voltage and the current. *Note 2:* Only effective power, *i.e.,* the actual power delivered to or consumed by the load, is expressed in watts. Volt-amperes reactive is properly expressed only in volt-amperes, never watts. *Note 3:* To maximize transmission efficiency, vars must be minimized by balancing capacitive and inductive loads, or by the addition of an appropriate capacitive or inductive reactance to the load.

volume. A portion of data, with its physical storage medium, that can be handled conveniently as a unit. *Note:* An example of a volume is a "floppy" diskette.

volume unit (vu). A unit of measurement of the power of an audio-frequency signal, as measured by a vu meter. *Note 1:* The vu meter is built and used in accordance with American National Standard C16.5-1942. *Note 2:* When using the vu meter to measure sine wave test tone power, 0 vu equals 0 dBm.

vox. An acoustoelectric transducer and a keying relay connected so that the keying relay is actuated when sound, or voice, energy above a certain threshold is sensed by the transducer. *Note:* A vox is used to eliminate the need for push-to-talk operation of a transmitter by using voice energy to turn on the transmitter. *Synonyms* voice operated relay circuit, voice operated transmit.

V reference point. The interface point in an ISDN environment between the line termination and the exchange termination.

VSB. *Abbreviation for* vestigial sideband. *See* vestigial sideband transmission.

V-series Recommendations. Sets of telecommunications protocols and interfaces defined by CCITT (now ITU-

T) Recommendations. *Note:* Some of the more common V.-series Recommendations are:

V.21. A CCITT Recommendation for modem communications over standard commercially available lines at 300 b/s. This protocol is generally not used in the United States.

V.22bis. A CCITT Recommendation for modem communications over standard commercially available voice-grade channels at 2,400 b/s and below.

V.32. A CCITT Recommendation for modem communications over standard commercially available voice-grade channels at 9.6 kb/s and below.

V.32bis. A CCITT Recommendation for modem communication over standard commercially available voice-grade channels at 14.4 kb/s and below.

V.34. An ITU-T Recommendation for modem communication over standard commercially available voice-grade channels at 28.8 kb/s and below.

V.42. A CCITT Recommendation for error correction on modem communications.

V.42bis. A CCITT Recommendation for data compression on a modem circuit.

V.FAST. A new CCITT Recommendation for high-speed modems currently under development.

VSWR. *Abbreviation for* voltage standing wave ratio.

vu. *Abbreviation for* volume unit.

wafer. A thin slice of semiconducting material, such as a silicon crystal, upon which microcircuits are constructed by diffusion and deposition of various materials. *Note:* Millions of individual circuit elements, constituting hundreds of microcircuits, may be constructed on a single wafer. The individual microcircuits are separated by scoring and breaking the wafer into individual chips ("dice").

WAIS. *Acronym for* Wide Area Information Servers. A distributed text searching system that uses the protocol standard ANS Z39.50 to search index databases on remote computers. *Note 1:* WAIS libraries are most often found on the Internet. *Note 2:* WAIS allows users to discover and access information resources on the network without regard to their physical location. *Note 3:* WAIS software uses the client-server model.

WAN. *Acronym for* wide area network.

wander. Relative to jitter and swim, long-term random variations of the significant instants of a digital signal from their ideal positions. *Note 1:* Wander variations are those that occur over a period greater than 1 s (second). *Note 2:* Jitter, swim, wander, and drift have increasing periods of variation in that order.

warm boot. *Synonym* warm restart (def. #2).

warm restart. 1. A sequence of operations that is performed to reset a previously running system, after an unintentional shutdown. *Synonym* warm start. 2. In computer operations, the restarting of equipment, after a sudden shutdown, that allows reuse of previously retained initialized input data, retained programs, and retained output queues. *Note 1:* A warm restart may be needed after a program failure. *Note 2:* A warm start or restart cannot occur if initial data, programs, and files are not retained after closedown. *Synonyms* hot boot, warm boot.

warm start. *Synonym* warm restart (def. #2).

Warner exemption. A statutory exemption pertaining to the acquisition of telecommunications systems that meet the exclusionary criteria of the Warner Amendment, Public Law 97-86, 1 December 1981, which is also known as the Brooks Bill. *Note:* Use of FTS2000 by U.S. Government agencies is mandatory when telecommunications are required. However, the Warner Amendment excludes the mandatory use of FTS2000 in instances related to maximum security.

WATS. *Acronym for* Wide Area Telephone Service.

wave equation. *See* Maxwell's equations.

waveform. The representation of a signal as a plot of amplitude versus time.

wavefront. The surface defined by the locus of points that have the same phase, *i.e.,* have the same path length from the source. *Note 1:* The wavefront is perpendicular to the ray that represents an electromagnetic wave. *Note 2:* The plane in which the electric and magnetic field vectors lie is tangential to the wavefront at every point. *Note 3:* The vector that represents the wavefront indicates the direction of propagation. *Note 4:* For parallel, *i.e.*, collimated, rays, the wavefront is plane. For rays diverging from a point, or converging toward a point, the wavefront is spherical. For rays with varying divergence or convergence, the wavefront has other shapes, such as ellipsoidal and paraboloidal, depending on the nature of the source.

waveguide. A material medium that confines and guides a propagating electromagnetic wave. *Note 1:* In the microwave regime, a waveguide normally consists of a hollow metallic conductor, usually rectangular, elliptical, or circular in cross section. This type of waveguide may, under certain conditions, contain a solid or gaseous dielectric material. *Note 2:* In the optical regime, a waveguide used as a long transmission line consists of a solid dielectric filament (optical fiber), usually circular in cross section. In integrated optical circuits an optical waveguide may consist of a thin dielectric film. *Note 3:* In the rf regime, ionized layers of the stratosphere and refractive surfaces of the troposphere may also act as a waveguide.

waveguide dispersion. *See* dispersion.

waveguide scattering. Scattering (other than material scattering) that is attributable to variations of geometry and refractive index profile of an optical fiber.

wave impedance. At a point in an electromagnetic wave, the ratio of the electric field strength to the magnetic field strength. *Note 1:* If the electric field strength is expressed in volts per meter and the magnetic field strength is expressed in ampere-turns per meter, the wave impedance will have the units of ohms. The wave impedance, Z, of an electromagnetic wave is given by

$$Z = \sqrt{\frac{\mu}{\varepsilon}} \ ,$$

where μ is the magnetic permeability and ε is the electric permittivity. For free space, these values are 4π x 10^{-7} H/m (henries per meter) and (1/36π) F/m (farads per meter), from which 120π, *i.e.*, 377, ohms is obtained. In dielectric materials, the wave impedance is 377/n, where *n* is the refractive index. *Note 2:* Although the ratio is called the wave impedance, it is also the impedance of the free space or the material medium.

wavelength. The distance between points of corresponding phase of two consecutive cycles of a wave. *Note:* The wavelength, λ, is related to the propagation velocity, *v*, and the frequency, *f*, by $\lambda = v/f$.

wavelength-division multiplexing (WDM). In optical fiber communications, any technique by which two or more optical signals having different wavelengths may be simultaneously transmitted in the same direction over one fiber, and then be separated by wavelength at the distant end.

wavelength stability. Of an optical source during a specified period, the maximum deviation of the peak wavelength from its mean value.

wave trap. A device used to exclude unwanted frequency components, such as noise or other interference, of a wave. *Note:* Traps are usually tunable to permit selection of unwanted or interfering signals.

WDM. *Abbreviation for* wavelength-division multiplexing.

weakly guiding fiber. An optical fiber in which the refractive index contrast is small (substantially less than 1%).

web browser. A user interface (usually graphical) to hypertext information on the World Wide Web.

weighting network. A network having a loss that varies with frequency in a predetermined manner, and which network is used for improving or correcting transmission characteristics, or for characterizing noise measurements.

whip antenna. A flexible rod antenna, usually between 1/10 and 5/8 wavelength long, supported on a base insulator.

white area. The area or population which does not receive interference-free primary service from an authorized AM station or does not receive a signal strength of at least 1 mV/m from an authorized FM station.

white facsimile transmission. 1. In an amplitude-modulated facsimile system, transmission in which the maximum transmitted power corresponds to the minimum density, *i.e.,* the white area, of the object. 2. In a frequency-modulated facsimile system, transmission in which the lowest transmitted frequency corresponds to the minimum density *i.e.,* the white area, of the object.

white noise. Noise having a frequency spectrum that is continuous and uniform over a specified frequency band. *Note:* White noise has equal power per hertz over the specified frequency band. *Synonym* additive white gaussian noise.

white pages. 1. A hard copy telephone directory listing of subscriber names, addresses, and telephone numbers. *Note: White pages* is associated with the residential subscriber listings in the standard directories distributed by the Bell System before divestiture. 2. An electronic information database that contains user names and their associated network addresses, in the manner of a telephone directory. *Note*: Electronic white pages usually contain additional information, such as office location, phone number, and mailstop.

white signal. In facsimile systems, the signal resulting from scanning a minimum-density area, *i.e.,* the white area, of the object.

who-are-you (WRU) character. A transmission-control character used for (a) switching on an answer-back unit in the station with which the connection has been established, (b) triggering the receiving unit to transmit an answer-back code to the terminal that transmitted the WRU signal, and (c) initiating a response that might include station identification, an indication of the type of equipment that is in service, and the status of the station. *Note 1:* The WRU signal corresponds to the 7-bit code assigned to the WRU. *Note 2:* The receiving unit may be a telegraph unit, data terminal equipment (DTE), or other unit. *Synonym* WRU signal.

Wide Area Information Servers (WAIS). *See* WAIS.

wide area network (WAN). A physical or logical network that provides data communications to a larger number of independent users than are usually served by a local area network (LAN) and is usually spread over a larger geographic area than that of a LAN. *Note 1:* WANs may include physical networks, such as Integrated Services Digital Networks (ISDNs), X.25 networks, and T1 networks. *Note 2:* A metropolitan area network (MAN) is a WAN that serves all the users in a metropolitan area. WANs may be nationwide or worldwide.

Wide Area Telephone Service (WATS). A toll service offering for customer dial-type telecommunications between a given customer [user] station and stations within specified geographic rate areas employing a single access line between the customer [user] location and the serving central office. Each access line may be arranged for either outward (OUT-WATS) or inward (IN-WATS) service, or both. *Note:* The offering is for fixed-rate inter- and intra-LATA services measured by zones and hours.

wideband. 1. The property of any communications facility, equipment, channel, or system in which the range of frequencies used for transmission is greater than 0.1 % of the midband frequency. *Note: "Wideband"* has many meanings depending upon application. *"Wideband"* is often used to distinguish it from *"narrowband,"* where both terms are subjectively defined relative to the implied context. 2. In communications security systems, a bandwidth exceeding that of a nominal 4-kHz telephone channel. 3. The property of a circuit that has a bandwidth wider than normal for the type of circuit, frequency of operation, or type of modulation. 4. In telephony, the property of a circuit that has a bandwidth greater than 4 kHz. 5. Pertaining to a signal that occupies a broad frequency spectrum. *Synonym* broadband.

wideband channel. A communication channel of a bandwidth equivalent to twelve or more voice-grade channels.

wideband modem. 1. A modem whose modulated output signal can have an essential frequency spectrum that is broader than that which can be wholly contained within, and faithfully transmitted through, a voice channel with a nominal 4-kHz bandwidth. 2. A modem whose bandwidth capability is greater than that of a narrowband modem.

wildcard character. 1. A character that may be substituted for any of a defined subset of all possible characters. *Note 1:* In high-frequency (HF) radio automatic link establishment, the wildcard character "?" may be substituted for any one of the 36 characters, "A" through "Z" and "0" through "9." *Note 2:* Whether the wildcard character represents a single character or a string of characters must be specified. 2. In computer (software) technology, a character that can be used to substitute for any other character or characters in a string. *Note:* The asterisk (*) usually substitutes as a wildcard character for any one or more of the ASCII characters, and the question mark (?) usually substitutes as a wildcard character for any one ASCII character.

wild-point detection. *Synonym* reasonableness check.

WIN. *Abbreviation for* WWMCCS Intercomputer Network.

window. 1. In fiber optics, a band of wavelengths at which an optical fiber is sufficiently transparent for practical use in communications applications. *Synonyms* spectral window, transmission window. *See* first window, second window, third window. 2. In imagery, a portion of a display surface in which display images pertaining to a particular application can be presented. *Note:* Different applications can be displayed simultaneously in different windows. 3. A period during which an event can occur, can be expected to occur, or is allowed to occur.

windowing. Sectioning of a video display area into two or more separate regions for the purpose of displaying images from different sources. *Note:* In windowing, one window could display data, another motion video from a remote site, and another, graphics.

wink. In telephone switching systems, a single supervisory pulse, *i.e.*, the momentary presence of, or interruption of, a supervisory signal. *Note:* An example of a wink is the momentary flash of a supervisory light on an attendant's switchboard.

wink pulsing. In telephone switching systems, recurring pulsing in which the off-condition is relatively short compared to the on-condition. *Note:* On key-operated telephone instruments, the hold position, *i.e.,* the hold condition, of a line is often indicated by wink pulsing the associated lamp at 120 pulses per minute. During 6% of the pulse period the lamp is off and 94% of the period the lamp is on, *i.e.,* 30 ms (milliseconds) off and 470 ms on.

wired radio frequency systems. Systems employing restricted radiation devices in which the radio frequency energy is conducted or guided along wires or in cables, including electric power and telephone lines.

wireless access mode. In personal communications service, interfacing with a network access point by means of a standardized air interface protocol without the use of a hardwired connection to the network.

wireless mobility management. In Personal Communications Service (PCS), the assigning and controlling of wireless links for terminal network connections. *Note:* Wireless mobility management provides an "alerting" function for call completion to a wireless terminal, monitors wireless link performance to determine when an automatic link transfer is required, and coordinates link transfers between wireless access interfaces.

wireless terminal. Any mobile terminal, mobile station, personal station, or personal terminal using non-fixed access to the network.

wireline common carrier. Common carriers [that] are in the business of providing landline local exchange telephone service.

word. A character string or a bit string considered to be an entity for some purpose. *Note:* In telegraph communications, six character intervals are defined as a word when computing traffic capacity in words per minute, which is computed by multiplying the data signaling rate in baud by 10 and dividing the resulting product by the number of unit intervals per character.

word length. The number of characters or bits in a word.

word processing. The use of a computer system to manipulate text. *Note:* Examples of word processing functions include entering, editing, rearranging, sorting, storing, retrieving, displaying, and printing text. *Synonym* text processing.

work space. In computers and data processing systems, the portion of main storage that is used by a computer program for the temporary storage of data.

work station. 1. In automated systems, such as computer, communications, and control systems, the input, output, display, and processing equipment that provides the operator-system interface. 2. A configuration of input, output, display, and processing equipment that constitutes a stand-alone system not requiring external access.

World Time. *Synonym* Coordinated Universal Time.

World Wide Web (WWW). An international, virtual-network-based information service composed of Internet host computers that provide on-line information in a specific hypertext format. *Note 1:* WWW servers provide hypertext metalanguage (HTML) formatted documents using the hypertext transfer protocol (HTTP). *Note 2:* Information on the WWW is accessed with a hypertext browser such as Mosaic, Viola, or Lynx. *Note 3:* No hierarchy exists in the WWW, and the same information may be found by many different approaches.

worst hour of the year. That hour of the year during which the median noise over any radio path is at a maximum. *Note:* This hour is considered to coincide with the hour during which the greatest transmission loss occurs.

W-profile fiber. *Synonym* doubly clad fiber.

wrapping. 1. In a network using dual counter-rotating ring architecture, reconfiguration to circumvent a failed link or node. 2. In open systems architecture, the use of a network to connect two other networks, thus providing an increased interaction capability between the two connected networks. *Note:* Recurring application of wrapping usually results in a hierarchical structure.

write. To make a permanent or transient recording of data in a storage device or on a data medium.

write cycle time. The minimum time interval between the starts of successive write cycles of a storage device that has separate reading and writing cycles.

write head. A magnetic head capable of writing only.

write protection label. *See* write-protect tab.

write-protect tab. A movable or removable tab, label, or other device, the presence or absence of which on the casing of a recording medium prevents writing on the medium. *Note:* An example of a write-protect tab is the sliding tab on a 3 ½-inch (8.85-cm) magnetic diskette of the type used in conjunction with desktop computers.

WRU signal. *Synonym* who-are-you (WRU) character.

WWMCCS. *Abbreviation for* Worldwide Military Command and Control System.

X-dimension of recorded spot. In facsimile systems, the effective recorded spot dimension measured in the direction of the recorded line. *Note 1:* By "effective recorded spot dimension" is meant the largest center-to-center spacing between recorded spots, which gives minimum peak-to-peak variation of density of the recorded line. *Note 2:* "X-dimension of recorded spot" implies that the facsimile equipment response to a constant density in the object (original) is a succession of discrete recorded spots.

X-dimension of scanning spot. In facsimile systems, the distance between the centers of adjacent scanning spots measured in the direction of the scanning line on the object. *Note:* The numerical value of the X-dimension of scanning spot depends upon the type of system.

xerographic recording. Recording enabled by the scanning action of an optical beam on a photoconducting surface on which an electrostatically charged latent image is developed with a resinous powder (toner).

XO. *Abbreviation for* crystal oscillator.

XOFF. An abbreviation for the ASCII transmission-control character meaning "Transmitter off."

XON. An abbreviation for the ASCII transmission-control character meaning "Transmitter on."

X.-series Recommendations. Sets of data telecommunications protocols and interfaces defined by CCITT Recommendations. *Note:* Some of the more common X.-series Recommendations are:

X.25. A CCITT Recommendation for public packet switched communications between a network user and the network itself.

X.75. A CCITT Recommendation for public packet switched communications between network hubs.

X.400. An addressing scheme for use with E-mail.

X.500. An addressing scheme for directory services.

XT. *Abbreviation for* crosstalk.

x-y mount. *Synonym* altazimuth mount.

Yagi antenna. A linear end-fire antenna, consisting of three or more half-wave elements (one driven, one reflector, and one or more directors). *Note 1:* A Yagi antenna offers very high directivity and gain. *Note 2:* The formal name for a *"Yagi antenna"* is *"Yagi-Uda array."*

Y-dimension of recorded spot. In facsimile systems, the distance between the centers of adjacent recorded spots on adjacent lines measured perpendicular to the recorded line.

Y-dimension of scanning spot. In facsimile systems, the distance between the centers of adjacent scanning spots on adjacent lines measured perpendicular to the scanning line on the object. *Note:* The numerical value of the Y-dimension of scanning spot depends upon the type of system.

Z. *Abbreviation for* Zulu time. *See* Coordinated Universal Time.

zero-bit insertion: A bit-stuffing technique used with bit-oriented protocols to ensure that six consecutive "1" bits never appear between the two flags that define the beginning and the ending of a transmission frame. *Note:* When five consecutive "1" bits occur in any part of the frame other than the beginning and ending flag, the sending station inserts an extra "0" bit. When the receiving station detects five "1" bits followed by a "0" bit, it removes the extra "0" bit, thereby restoring the bit stream to its original value.

zero dBm transmission level point (0 dBm TLP). In a communication system, a point at which the reference level is 1 mW, *i.e.,* 0 dBm. *Note:* The actual power level of the communications traffic is not necessarily 0 dBm. It is usually below the reference level. The reference is for system design and test purposes. *Synonym* zero transmission level point.

zero-dispersion slope. In a single-mode optical fiber, the rate of change of dispersion, with respect to wavelength, at the fiber's zero-dispersion wavelength. *Note 1:* In silica-based optical fibers, the zero-dispersion wavelength occurs at approximately 1.3 µm, but this wavelength may be shifted toward the minimum-loss window by the addition of dopants to the fiber material during manufacture. *Note 2:* Doubly and quadruply clad single-mode fibers have two zero-dispersion points, and thus two zero-dispersion slopes.

zero-dispersion wavelength. 1. In a single-mode optical fiber, the wavelength or wavelengths at which material dispersion and waveguide dispersion cancel one another. *Note:* In all silica-based optical fibers, minimum material dispersion occurs naturally at a wavelength of approximately 1.3 μm. Single-mode fibers may be made of silica-based glasses containing dopants that shift the material-dispersion wavelength, and thus, the zero-dispersion wavelength, toward the minimum-loss window at approximately 1.55 μm. The engineering tradeoff is a slight increase in the minimum attenuation coefficient. 2. Loosely, in a multimode optical fiber, the wavelength at which material dispersion is minimum, *i.e.*, essentially zero. *Synonym* minimum-dispersion wavelength.

zero dispersion window. *Synonym* minimum dispersion window.

zerofill. To fill unused storage locations with the representation of the character denoting "0".

zero-level decoder. A decoder that yields an analog level of 0 dBm at its output when the input is the digital mailed signal. *Note:* The signal is a 1-kHz sine wave.

zero suppression. The elimination of nonsignificant zeros from a numeral.

0TLP. *Abbreviation for* zero transmission level point.

zero transmission level point (0TLP). *Synonym* zero dBm transmission level point.

zip-cord. In optical communications, a two-fiber cable consisting essentially of two single-fiber cables having their jackets conjoined by a strip of jacket material. *Note 1:* This name is borrowed from electrical terminology referring to lamp cord. As with lamp cord, optical zip-cord may be easily furcated by slitting or tearing the two jackets apart, permitting the installation of optical connectors. *Note 2:* Zip-cord cables include both loose-buffer and tight-buffer designs.

zone. *See* communications zone, Fresnel zone, skip zone.

zone of silence. *Synonym* skip zone.

Z Time. *Abbreviation for* Zulu Time. *See* Coordinated Universal Time.

Zulu Time (Z). *Synonym* Coordinated Universal Time. *Formerly a synonym for* Greenwich Mean Time.